Springer Atmospheric Sciences

More information about this series at http://www.springer.com/series/10176

Amujuri Mary Selvam

Self-organized Criticality and Predictability in Atmospheric Flows

The Quantum World of Clouds and Rain

Springer

Amujuri Mary Selvam
Ministry of Earth Sciences,
Government of India
Retired from IITM
Pune
India

ISSN 2194-5217 ISSN 2194-5225 (electronic)
Springer Atmospheric Sciences
ISBN 978-3-319-54545-5 ISBN 978-3-319-54546-2 (eBook)
DOI 10.1007/978-3-319-54546-2

Library of Congress Control Number: 2017938286

Printed on acid-free paper

This Springer imprint is published by Springer Nature
The registered company is Springer International Publishing AG
The registered company address is: Gewerbestrasse 11, 6330 Cham, Switzerland

To my husband Murty
and Grandchildren Rohan and Tara

Preface

Nonlinear dynamical systems in nature, such as atmospheric flows, exhibit complex spatial patterns, e.g. cloud geometry, that lack a characteristic (single) length scale concomitant with temporal fluctuations that lack a single timescale. The mathematical concept of 'fractals' introduced by Mandelbrot (1977) provides powerful tools for describing and quantifying the universal symmetry of self-similarity (Schroeder 1991) underlying the seemingly irregular complex geometrical shapes and temporal fluctuations.

Spatially extended fractal objects in nature support fluctuations of dynamical processes on all timescales. The power spectra of such broadband fluctuations exhibit an inverse power law of form $1/f^{\alpha}$, where f is the frequency and α the exponent. In general, α decreases with f and approaches 1 for low frequencies. Such spectra, described as $1/f$ or $1/f$-like spectra of temporal fluctuations, are ubiquitous to dynamical systems in nature. The frequency range over which α is constant therefore exhibits self-similarity or scale invariance in temporal fluctuations, i.e. the fluctuations are fractals in time. The intensity or variance of longer and shorter period fluctuations is mutually related by a scale factor alone, independent of the nature of dynamical processes. The fluctuations exhibit long-range temporal correlations. Also, temporal fluctuations exhibit multifractal structure because α varies for different ranges of frequency f. $1/f$ power law would seem natural, and white noise (flat distribution) would be the subject of involved investigation.

Until very recently (1988), fractal geometry to the spatial pattern and fractal fluctuations in time of dynamical processes of the same extended dynamical system were treated as two disparate multidisciplinary fields of research. The long-range spatiotemporal correlations underlying spatial and temporal power law behaviour of dynamical systems were identified as a unified manifestation of self-organized criticality in 1988 (Bak et al. 1987, 1988).

The unifying concept of self-organized criticality underlying fractals, self-similar scaling, broadband frequency spectra, and inverse power law distributions offers a new and powerful means of describing certain basic aspects of spatial form and dynamical (temporal) processes of a dynamical system. The systems in which self-organized criticality is observed range from the physical to the biological to the social. Rapid advances in applications of these new concepts have been made, particularly in the field of physiology and medicine. It is now recognized that fractal architecture to the spatial pattern serves as robust, stable structures for the regulation and maintenance of vital functions of lungs, heart, liver, kidneys, brain, etc. Self-similar fractal growth pattern in the plant kingdom gives rise to the observed beautiful phyllotactic patterns, i.e. the elaborate patterns of fruits in the sunflower capitulum, florets in the capitulum of daisy, and scales on a pineapple and on a pinecone. Phyllotactic patterns incorporate with mathematical precision the Fibonacci mathematical series, where each term is the sum of two previous terms and the ratio of each term to the previous term approaches the golden ratio $\tau = (1 + \sqrt{5})/2 \approx 1.618$. Such patterns, while pleasing to the eye, combine maximum packing efficiency while preserving the shape for different sizes of a particular species, such as daisy flowers of different sizes.

Atmospheric flows exhibit self-organized criticality, i.e. long-range correlations in space and time manifested as fractal geometry to the spatial pattern concomitant with inverse power law form for fluctuations of meteorological parameters such as temperature and pressure. Traditional meteorological theory cannot explain satisfactorily the observed self-similar space-time structure of atmospheric flows. A recently developed general systems theory for fractal space-time fluctuations shows that the larger-scale fluctuation can be visualized to emerge from the space-time averaging of enclosed small-scale fluctuations, thereby generating a hierarchy of self-similar fluctuations manifested as the observed eddy continuum in power spectral analyses of fractal fluctuations. The interconnected network of eddy circulations responds as a unified whole to local perturbation such as global-scale response to El Nino events. The general systems theory model predicts inverse power law form incorporating the golden ratio τ for the distribution of space-time fluctuation pattern and also for the power (variance) spectra of the fluctuations. Since the probability distributions of amplitude and variance are the same, atmospheric flows exhibit quantum-like chaos. Long-range correlations inherent in power law distributions of fluctuations are identified as non-local connection or entanglement exhibited by quantum systems such as electron or photon. The predicted distribution is close to the Gaussian distribution for small-scale fluctuations, but exhibits *fat long tail* for large-scale fluctuations. Universal inverse power law for fractal fluctuations rules out unambiguously linear secular trends in climate parameters. Energy input into the atmospheric eddy continuum, either natural or man-made, will result in the enhancement of fluctuations of all scales manifested immediately in intensification of high-frequency fluctuations such as the Quasi-Biennial Oscillation (QBO) and El Nino-Southern Oscillation (ENSO) cycles.

Chapter 1 gives a brief review of the science of nonlinear dynamics and chaos with applications in meteorology and atmospheric physics. Atmospheric flows, an example of turbulent fluid flows, exhibit fractal fluctuations of all space-time scales ranging from turbulence scale of mm-sec to climate scales of thousands of kilometres-years and may be visualized as a nested continuum of weather cycles or periodicities, the smaller cycles existing as intrinsic fine structure of the larger cycles. The power spectra of fractal fluctuations exhibit inverse power law form signifying long-range correlations identified as self-organized criticality and are ubiquitous to dynamical systems in nature and are manifested as sensitive dependence on initial condition or 'deterministic chaos' in finite precision computer realizations of nonlinear mathematical models of real-world dynamical systems such as atmospheric flows. Though the self-similar nature of atmospheric flows has been widely documented and discussed during the last three to four decades, the exact physical mechanism is not yet identified. There now exists an urgent need to develop and incorporate basic physical concepts of nonlinear dynamics and chaos into classical meteorological theory for more realistic simulation and prediction of weather and climate. A historical review of nonlinear dynamics and chaos in meteorology and atmospheric physics is summarized in this chapter.

Chapter 2 gives a review of noise or random fluctuations in physical systems. 'Noise' or random fluctuations characterize all physical systems in nature ranging from biology, botany, physiology, meteorology, astronomy, etc. The apparently irregular or chaotic fluctuations were considered as 'noise' in all fields except in astronomy where the fluctuations from astronomical sources were referred to as signal. Noise and fluctuation have been a field of study since 1826 with the study of Brownian motion which indirectly confirmed the existence of atoms and molecules. The measured characteristics of noise contain recognizable patterns or signal and convey useful information about the system. Statistical data analysis techniques are used to extract the signal, i.e. recognizable patterns in the apparently random fluctuations of physical systems. The analysis of data sets and broad quantification in terms of probabilities belongs to the field of statistics. Early attempts resulted in the identification of the following two quantitative (mathematical) distributions which approximately fit data sets from a wide range of scientific and other disciplines of study. The first is the well-known statistical normal distribution, and the second is the power law distribution associated with the recently identified 'fractals' or self-similar characteristic of data sets in general. Abraham de Moivre, an eighteenth century statistician and consultant to gamblers, made the first-recorded discovery of the normal curve of error (or the bell curve because of its shape) in 1733. The importance of the normal curve stems primarily from the fact that the distributions of many natural phenomena are at least approximately normally distributed. This normal distribution concept underlies how we analyse experimental data over the last two hundred years. Most quantitative research involves the use of statistical methods presuming *independence* among data points and Gaussian 'normal' distributions. The Gaussian distribution is reliably characterized by its stable mean and finite variance. Normal distributions place a trivial amount of probability far from the mean, and hence, the mean is representative of most observations. Even the largest deviations, which are exceptionally rare, are still only about a factor of two from the mean in either direction and are well characterized by quoting a simple standard deviation. However, apparently rare real-life catastrophic events such as major earth quakes, stock market crashes, and heavy rainfall events, occur more frequently than indicated by the normal curve; i.e., they exhibit a probability distribution with a *fat tail*. Fat tails indicate a power law pattern and interdependence.

The 'tails' of a power law curve—the regions to either side that correspond to large fluctuations—fall off very slowly in comparison with those of the bell curve. The normal distribution is therefore an inadequate model for extreme departures from the mean. For well over a century, evidence had been mounting that real-world behaviour, in particular, behaviour of *systems*, whether natural, social, economic, or financial does not follow normal distribution characteristics. There is increased evidence for non-normality in real-world settings and in its place an alternative distribution, namely the power law distribution, is shown to be exhibited by real-world systems in all fields of science and other areas of human interest. In this chapter, the following are discussed: (i) a brief history of the two chief quantitative methods of statistical data analysis, namely the statistical normal distribution and the power law distribution; (ii) the association of power law distributions with complex systems, scale invariance, self-similarity, fractals, $1/f$ noise, long-term memory, phase transitions, critical phenomena, and self-organized criticality; (iii) current status of power law distributions; (iv) power law relations (bivariate) and power law (probability) distributions; (v) allometric scaling and fractals; (vi) fractals and the golden section in plant growth; (vii) turbulent fluid flow structure, fractals, and the golden ratio ($\approx$1.618); (viii) fractal space-time and the golden ratio; (ix) power law (probability) distributions in the meteorological parameter precipitation, temperature, quaternary ice volume fluctuations, and atmospheric pollution; and (x) general systems theory model for self-organized criticality in atmospheric flows with universal quantification for power law distribution in terms of the golden ratio.

In Chap. 3, a general systems theory model for atmospheric flows is presented and is shown that self-organized criticality (SOC) is a signature of quantum-like chaos.

Atmospheric flows exhibit long-range spatiotemporal correlations manifested as the fractal geometry to the global cloud cover pattern concomitant with inverse power law form for power spectra of temporal fluctuations on all space-time scales ranging from turbulence (centimetres-seconds) to climate (kilometres-years). Long-range spatiotemporal correlations are ubiquitous to dynamical systems in nature and are identified as signatures of *self-organized criticality*. Standard models in meteorological theory cannot explain satisfactorily the observed self-organized criticality in atmospheric flows. Mathematical models for simulation and prediction of atmospheric flows are nonlinear and do not possess analytical solutions. Finite precision computer realizations of nonlinear models give unrealistic solutions because of *deterministic chaos*, a direct consequence of round-off error growth in iterative numerical computations. Recent studies show that round-off error doubles on an average for each iteration of iterative computations. Round-off error propagates to the mainstream computation and gives unrealistic solutions in numerical weather prediction (NWP) and climate models, which incorporate thousands of iterative computations in long-term numerical integration schemes. A general systems theory model for atmospheric flows developed by the author predicts the observed *self-organized criticality* as intrinsic to quantum-like chaos in flow dynamics. The model provides universal quantification for *self-organized criticality* in terms of the golden ratio τ (≈1.618). Model predictions are in agreement with a majority of observed spectra of time series of several standard meteorological and climatological data sets representative of disparate climatic regimes. Universal spectrum for natural climate variability rules out linear trends. Man-made greenhouse gas-related atmospheric warming would result in intensification of natural climate variability, seen immediately in high-frequency fluctuations such as QBO and ENSO and even shorter timescales. Model concepts and results of analyses are discussed with reference to possible prediction of climate change. Model concepts, if correct, rule out unambiguously linear trends in climate. Climate change will only be manifested as increase or decrease in the natural variability. However, more stringent tests of model concepts and predictions are required before applications to such an important issue as *climate change*. The cell dynamical system model for atmospheric flows is a general systems theory applicable, in general, to all dynamical systems in other fields of science, such as *number theory*, *biology*, *physics*, and *botany*.

In Chap. 4, it is shown that the distribution for month-wise temperature and rainfall in the UK region follows the universal inverse power law distribution predicted by the general systems theory model discussed in Chap. 2. S. Lovejoy and his group at the McGill University, Canada, have done pioneering work during the last three decades to show conclusively that meteorological parameters, such as temperature, rainfall, and pressure, exhibit self-similar space-time fractal fluctuations generic to dynamical systems in nature such as fluid flows, spread of forest fires, and earthquakes. The power spectra of fractal fluctuations display inverse power law form signifying long-range correlations. A general systems theory model (Chap. 2) predicts universal inverse power law form incorporating the golden ratio for the fractal fluctuations. The model-predicted distribution was compared with observed distribution of fractal fluctuations of all size scales (small, large, and extreme values) in the historic month-wise temperature (maximum and minimum) and total rainfall for the four stations Oxford, Armagh, Durham, and Stornoway in the UK region, for data periods ranging from 92 years to 160 years. For each parameter, the two cumulative probability distributions, namely *cmax* and *cmin*, starting from, respectively, maximum and minimum data value were used. The results of the study show that (i) temperature distributions (maximum and minimum) follow model-predicted distribution except for Stornowy, minimum temperature *cmin*; (ii) rainfall distribution for *cmin* follows model-predicted distribution for all the four stations; and (iii) rainfall distribution for *cmax* follows model-predicted distribution for the two stations Armagh and Stornoway. The present study suggests that fractal fluctuations result from the superimposition of eddy continuum fluctuations.

In Chap. 5, it is shown that signatures of universal characteristics of fractal fluctuations are seen in global mean monthly temperature anomalies. Self-similar space-time fractal fluctuations are generic to dynamical systems in nature such as atmospheric flows, heartbeat patterns, and population dynamics. The physics of the long-range correlations intrinsic to fractal fluctuations is not completely understood. It is important to quantify the physics underlying the irregular fractal fluctuations for prediction of space-time evolution of dynamical systems. A general systems theory model (Chap. 2) for fractals visualizes the emergence of successively larger-scale fluctuations resulting from the space-time integration of enclosed small-scale fluctuations. The theoretical model predictions are as follows: (i) the probability distribution and the power spectrum for fractal fluctuations are the same inverse power law function incorporating the *golden ratio*. (ii) The predicted distribution is close to the Gaussian distribution for small-scale fluctuations but exhibits *fat long tail* for large-scale fluctuations with higher probability of occurrence than predicted by Gaussian distribution. (iii) Since the power spectrum (variance, i.e., square of eddy amplitude) also represents the probability densities as in the case of quantum systems such as the electron or photon, fractal fluctuations exhibit quantum-like chaos. (iv) The *fine structure constant* for spectrum of fractal fluctuations is a function of the golden ratio and is analogous to atomic spectra equal to about 1/137. Global gridded time series data sets of monthly mean temperatures for the period 1880–2007/2008 were analysed. The data sets and the corresponding power spectra exhibit distributions close to the model-predicted inverse power law distribution. The model-predicted and observed universal spectrum for interannual variability rules out linear secular trends in global monthly mean temperatures. Global warming either man-made or natural results in intensification of fluctuations of all scales and manifests immediately in high-frequency fluctuations.

Amujuri Mary Selvam
Deputy Director (Retired)
Address (Res): B1 Aradhana, 42/2A Shivajinagar,
Pune 411005, India

References

Bak, P., Tang, C., Wiesenfeld, K.: Self-organized criticality: an explanation of *1/f* noise. Phys. Rev. Lett. **59**, 381–384 (1987)

Bak, P.C., Tang, C., Wiesenfeld, K.: Self-organized criticality. Phys. Rev. A. **38**, 364–374 (1988)

Mandelbrot, B.B.: Fractals: Form, Chance and Dimension. W. H. Free-man and Co., San Francisco (1977)

Schroeder, M.: Fractals, Chaos and Power-Laws. W. H. Freeman and Co., San Francisco (1991)

About the Book

Atmospheric flows exhibit self-organized criticality, i.e. long-range correlations in space and time manifested as fractal geometry to the spatial pattern concomitant with inverse power law form for fluctuations of meteorological parameters such as temperature and pressure. Traditional meteorological theory cannot explain satisfactorily the observed self-similar space-time structure of atmospheric flows. A recently developed general systems theory for fractal space-time fluctuations shows that the larger-scale fluctuation can be visualized to emerge from the space-time averaging of enclosed small-scale fluctuations, thereby generating a hierarchy of self-similar fluctuations manifested as the observed eddy continuum in power spectral analyses of fractal fluctuations. The interconnected network of eddy circulations responds as a unified whole to local perturbation such as global-scale response to El Nino events. The general systems theory model predicts inverse power law form incorporating the golden mean τ for the distribution of space-time fluctuation pattern and also for the power (variance) spectra of the fluctuations. Since the probability distributions of amplitude and variance are the same, atmospheric flows exhibit quantum-like chaos. Long-range correlations inherent to power law distributions of fluctuations are identified as non-local connection or entanglement exhibited by quantum systems such as electron or photon. The predicted distribution is close to the Gaussian distribution for small-scale fluctuations, but exhibits *fat long tail* for large-scale fluctuations. Universal inverse power law for fractal fluctuations rules out unambiguously linear secular trends in climate parameters. Energy input into the atmospheric eddy continuum, either natural or man-made, will result in the enhancement of fluctuations of all scales manifested immediately in intensification of high-frequency fluctuations such as the Quasi-Biennial Oscillation and El Nino-Southern Oscillation cycles.

This monograph will provide background reading for many postgraduate students of atmospheric physics, environmental sciences, and physics and scientists working in the field of the topic of the Monograph and also in the multidisciplinary field of nonlinear dynamics and chaos.

The monograph is based on substantial research work in atmospheric physics carried out by the author during the past two decades and aims to promote understanding of some of the basic mathematical and scientific issues in the subjects relating to climate dynamics, chaos, and quantum mechanics.

Contents

About the Author

Amujuri Mary Selvam Obtained Master's Degree in physics from the University of Madras, India, in 1960 and Ph.D. in physics from the University of Pune in 1970. She worked at India Meteorological Department (IMD) from 1962 to 1966. She joined Indian Institute of Tropical Meteorology (IITM), Department of Earth Sciences, Government of India, Pune, India, in 1966 and retired in 1999 as Deputy Director, Physical Meteorology and Aerology Division. She carried out extensive research work in atmospheric sciences and published several research papers in national and international journals and books entitled 'Chaotic Climate Dynamics', 2007, Luniver Press, UK., and 'Rain Formation in Warm Clouds: General Systems Theory', 2015, SpringerBriefs in Meteorology, Springer, Germany.

The author acknowledges the support and encouragement received from her colleagues and seniors at the IMD and IITM.

Web sites: http://amselvam.webs.com; http://amselvam.tripod.com; http://www.geocities.ws/amselvam.

About the Author

Chapter 1
Nonlinear Dynamics and Chaos: Applications in Meteorology and Atmospheric Physics

Abstract Atmospheric flows, an example of turbulent fluid flows, exhibit fractal fluctuations of all space-time scales ranging from turbulence scale of mm-sec to climate scales of thousands of kilometres/years and may be visualized as a nested continuum of weather cycles or periodicities, the smaller cycles existing as intrinsic fine structure of the larger cycles. The power spectra of fractal fluctuations exhibit inverse power-law form signifying long-range correlations identified as self-organized criticality and are ubiquitous to dynamical systems in nature and is manifested as sensitive dependence on initial condition or 'deterministic chaos' in finite precision computer realizations of nonlinear mathematical models of real-world dynamical systems such as atmospheric flows. Though the self-similar nature of atmospheric flows have been widely documented and discussed during the last three to four decades, the exact physical mechanism is not yet identified. There now exists an urgent need to develop and incorporate basic physical concepts of nonlinear dynamics and chaos into classical meteorological theory for more realistic simulation and prediction of weather and climate. A historical review of nonlinear dynamics and chaos in meteorology and atmospheric physics is summarized in this chapter.

Keywords Nonlinear dynamics and chaos · Weather and climate prediction · Fractals · Self-organized criticality · Long-range correlations · Inverse power law

1.1 Introduction

The history of nonlinear dynamics and chaos begins with the brilliant original contribution of Poincare' (1890) and specifically the derivation of the Poincare' map, and later with the advent of digital computers, of the findings of Ueda (1979) and Feigenbaum (1978) of the explosive and period-doubling routes to chaos, respectively. These findings, later complemented by the mathematics of global bifurcation theory (Guckenheimer and Holmes 1983) and analysis of observed chaotic data (Abarbanel 1996) set the stage for the well-established theory which today is the new science-motivating applications to atmospheric science.

A.M. Selvam, *Self-organized Criticality and Predictability in Atmospheric Flows*, Springer Atmospheric Sciences, DOI 10.1007/978-3-319-54546-2_1

Atmospheric flows, an example of turbulent fluid flows exhibits self-similar fractal fluctuations of all space-timescales ranging from turbulence scale of mm-sec to climate scales of kms/years and may be visualized as a nested continuum of cycles or periodicities, the smaller cycles existing as fine-scale structure of larger cycles. Fractal fluctuations are ubiquitous to dynamical systems in nature such as river flows, heart beat patterns, population dynamics, computer realizations of nonlinear mathematical models of dynamical systems, etc., and has been identified in all areas of science and human interest (Mandelbrot 1977). The power spectra of fractal fluctuations exhibit inverse power-law form $f^{-\alpha}$ where f is the frequency and α, the exponent. The frequency range over which α is a constant exhibits self-similarity or scale invariance, i.e. the fluctuation intensity (variance or amplitude squared) is a function of α alone and is independent of any other intrinsic property of the dynamical system such as its physical, chemical or any other characteristic. Scale invariance of space-time fluctuations of dynamical systems signifies long-range correlations or non-local connections and is identified as self-organized criticality (Bak et al. 1988). The physics of the observed universal characteristics of fractal fluctuations indicate a common physical mechanism governing the space-time evolution of dynamical systems. Therefore a general systems theory (Peacocke 1989; Klir 1993; Jean 1994) where the model concepts are independent of the exact details, such as the chemical, physical, physiological, etc., properties of the dynamical systems will be applicable to dynamical systems in nature. Identification of the physics of self-organized criticality will enable quantification of the space-time growth pattern of dynamical systems such as atmospheric flows for predictability of future evolution of weather patterns. In this paper, a review is given on the current status of application of recently developed concepts in the new multidisciplinary science of nonlinear dynamics and chaos in meteorology and atmospheric physics. The paper is organized as follows. A brief history of the new science of nonlinear dynamics and chaos is given in Sect. 1.2. The identification of fractals and self-organized criticality in meteorology and atmospheric physics is summarized in Sect. 1.3. The current status of applications of nonlinear dynamics and chaos for weather prediction is given in Sect. 1.4. Discussions and conclusions are given in Sect. 1.5.

1.2 New Science of Nonlinear Dynamics and Chaos

1.2.1 Dynamical Systems and Fractal Space-Time Fluctuations

Dynamical systems in nature, i.e. systems that change with time, such as fluid flows, heartbeat patterns, spread of infectious diseases, etc., exhibit nonlinear (unpredictable) fluctuations. Conventional mathematical and statistical theories deal only with linear systems and the exact quantification and description of nonlinear fluctuations was not possible till the identification in the 1970s by Mandelbrot (1977, 1983), of the universal symmetry of *self-similarity*, i.e. *fractal geometry* underlying the seemingly irregular fluctuations in space and time (Schroeder 1991; Stanley 1995). Fractals, as the name implies, describe non-Euclidean objects generic to nature such as tree roots, tree branches, river basins, etc., which occupy only a part (fraction) of the traditional (Euclidean) three or two dimensions (Mayer-Kress 1988). The study of self-similar space-time fluctuations generic to dynamical systems, now (since 1980s), belongs to the newly emerging multidisciplinary science of *nonlinear dynamics and chaos* (Gleick 1987) and deals with unified concepts for fundamental aspects intrinsic to the complex (nonlinear) and apparently random (chaotic) space-time structures found in nature. Scientific community at large will derive immense benefit in terms of new insights and development of powerful analytical techniques in this multidisciplinary approach to quantify basic similarities in form and function in disparate contexts ranging from the microscopic to the macroscopic scale.

The apparently random, noisy or irregular space-time signals (patterns) of a dynamical system, however, exhibit qualitative similarity in pattern geometry on all scales and are therefore correlated. In general, the spatiotemporal evolution of dynamical systems trace a zigzag (jagged) pattern of alternating increase and decrease, associated with bifurcation or branching on all scales of space and time, generating wrinkled or folded surfaces in three dimensions. Representative examples for time series of some meteorological parameters are shown in Fig. 1.1. Physical, chemical, biological and other dynamical systems exhibit similar universal irregular space-time fluctuations. A fascinating aspect of patterns in nature is that many of them have a universal character (Dennin et al. 1996).

Irregular space-time fluctuations associated with basic bifurcation or branching geometry of wrinkles or folds on all scales is associated with the symmetry of self-similarity under scales transformation or just self-similarity (Liu 1992). A symmetry principle is simply a statement that something looks the same from certain different points of view. Such symmetries are often called principles of invariance (Weinberg 1993). The fundamental similarity or universality in the basic geometric structure, namely, irregularity, was identified as *fractal* in the late 1970s by Mandelbrot (1977, 1983). *Fractal* geometry is ubiquitous in nature, the fine structure on all scales being the optimum design for sustenance and growth of large-scale complex systems comprised of an integrated network of subunits. The branching architecture of river tributaries, bronchial tree, tree branches, lightning discharge, etc., serves to collect/disperse fluids over a maximum surface area within a minimum volume. Fine-scale fluctuations help efficient mixing of fluids such as pollution dispersion in the atmosphere.

The basic similarity in the branching form underlying the individual leaf and the tree as a whole was identified more than three centuries ago in botany (Arber 1950). The importance of scaling concepts were recognized nearly a century ago in biology and botany where the dependence of a property y on size x is usually expressed by the allometric equation $y = ax^b$ where a and b are constants (Thompson 1963; Strathmann 1990; Jean 1994; Stanley et al. 1996b). This type of

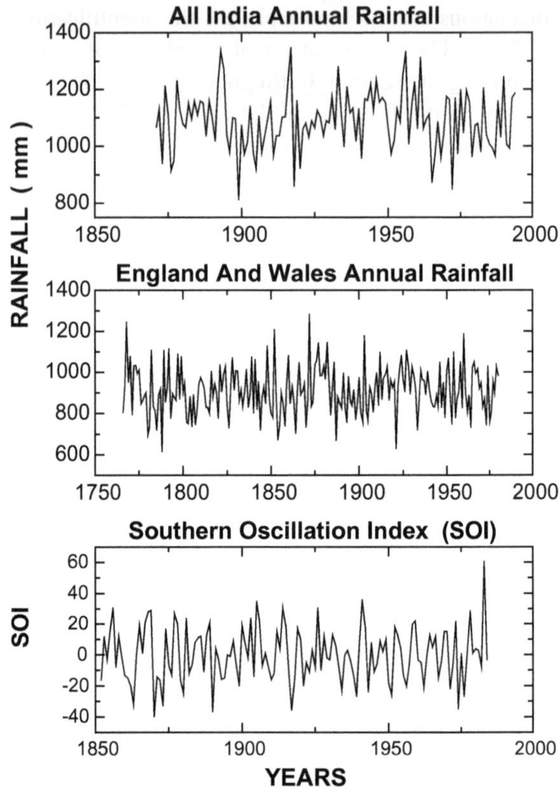

Fig. 1.1 Time series data of some of the meteorological parameters are shown as representative examples for irregular (zigzag) fluctuations (temporal) generic to dynamical systems in nature

scaling implies a hierarchy of substructures and was used by *D'Arcy Thompson* for scaling anatomical structures, for example, how proportions tend to vary as an animal grows in size (West 1990a, b). D'Arcy Thompson (1963, first published in 1917) in his book *On Growth and Form* has dealt extensively with similitude principle for biological modelling. Rapid advances have been made in recent years in the fields of biology and medicine in the application of scaling (*fractal*) concepts for description and quantification of physiological systems and their functions (Goldberger et al. 1990; West 1990a, b; Deering and West 1992; Skinner 1994; Stanley et al. 1996b). In meteorological theory, the concept of self-similar fluctuations was introduced in the description of turbulent flows by Richardson (1965, originally published in 1922), Kolmogorov (1941, 1962), Mandelbrot (1975), Kadanoff (1996) and others (see Monin and Yaglom 1975 for a review).

1.2.2 Fractals in Pure Mathematics

Irregular (wrinkled) patterns are often described by functions that are continuous but not differentiable. Till the late 1800s, pure mathematics dealt mostly with functions, which are differentiable everywhere such as the circle or ellipse. Pioneers in the study of functions which are continuous everywhere but without tangents are Karl Weierstrauss (1815–1897) who presented the Weierstrauss function in 1872, George Cantor (1845–1918) who provided the Cantor set in 1883 and Helge Van Koch (1906) who first constructed the snowflake curve (Deering and West 1992). A representative example, the Koch's curve is shown in Fig. 1.2.

Jagged boundaries represented by these functions are more common in nature than the special case of curves with tangents, such as the circle. However, real-world geometrical structures were not associated with these functions till a long time after their discovery. Continuous functions which are not differentiable anywhere represent an infinite number of zigzags between any two points. The length between any two points on the curve is infinity, yet the area bounded by the curve is finite. These 'monster curves' which were outside the domain of pure mathematics were ignored as a field for study by many prominent mathematicians till the late 1800s.

The non-Euclidean geometry of the 'monster curve' was quantified in terms of the similarity dimension by Hausdorff (1919). His idea was based on scaling, which means measuring the same object with different units of measurement. Any detail smaller than the unit of measurement is discarded. The jagged 'monster curves' have fractional (noninteger) dimensions. The word *fractal* was coined by Mandelbrot (1977) as a generic name for such objects as *Koch*'s snowflake, which possess fractional *Hausdorff* dimension. Besicovitch (1929) was a second major figure who had developed the background for the concept of fractional dimension.

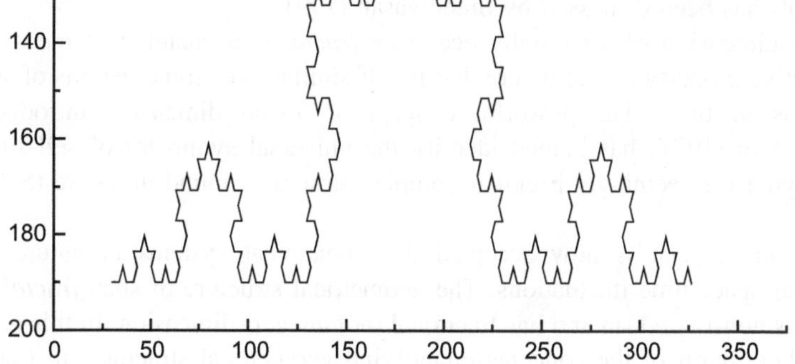

Fig. 1.2 The Koch's curve as a representative example for mathematical functions which are continuous everywhere but not differentiable anywhere, i.e. tangents cannot be drawn anywhere on the jagged boundary

Some of the earlier studies on applications of scaling concepts are given in the following. The question of scaling and the paradigm of *fractals*, i.e. when can a part have the same properties as the whole was addressed in the 1920s and 1930s by Levy (1937) who was concerned with the question of when a sum of identically distributed random variables has the same probability distribution as any one of the terms in the sum (Shlesinger et al. 1987). The length of a *fractal* object, e.g., the coastline increases with decrease in the length of yardstick used for the measurement. Richardson (1960) came close to the concept of *fractals* when he noted that the estimated length of an irregular coastline or boundary $B(l)$, where l is the measuring unit is given by $B(l) = B_o l^{1-d}$ where B_o is a constant with dimension of length and d is the *fractal* dimension greater than 1 but less than 2 for the jagged coastline (West 1990a, b). One of the oldest scaling laws in geophysics is the *Omori* law (Omori 1895). This law describes the temporal distribution of the number of after-shocks, which occur after a larger earthquake (i.e. mainshock) by a scaling relationship. The other basic empirical seismological law, the *Gutenberg–Richter* law (Gutenberg and Richter 1944) is also a scaling relationship, and relates intensity to its probability of occurrence (Hooge et al. 1994). The power law is a distinctive experimental signature seen in a wide variety of complex systems. In economics, it goes by the name of 'fat tails', in physics it is referred to as 'critical fluctuations', in computer science and biology it is 'the edge of chaos', and in demographics it is called Zipf's law (Newman 2000).

The *fractal* dimension D in general for length scale R may be given as

$$D = \frac{\mathrm{d}\,\ln M}{\mathrm{d}\,\ln R},$$

where M is the mass contained within a distance R from a point in the extended object. A constant value for D implies uniform stretching on logarithmic scale, resulting in large-scale structures which preserve their original geometrical shape. Objects in nature are in general *multifractals*, i.e. the *fractal* dimension D varies with the length scale R. The *multifractal* nature of fluid turbulence and scaling concepts has been discussed by Sreenivasan (1991).

The dimension of a naturally occurring *fractal* is a quantitative measure of a qualitative property of a structure that is self-similar over some regions of space or intervals of time. The powerful concept of fractal dimension introduced by Mandelbrot (1977) has helped identify the universal symmetry of self-similarity underlying the seemingly irregular complex structures found in nature (Schroeder 1991).

In summary, it is now accepted that dynamical systems in nature exhibit irregular space-time fluctuations. The geometrical structure of such *fractal* fluctuations is non-Euclidean and has fractional (noninteger) dimension. In this context a brief description of the concepts quantifying geometrical structures in traditional mathematics is given in the following. Classical Euclidean geometry deals only with regular objects such as *point*, *line*, *square* and *cube* in terms of integer dimensions *zero*, *one*, *two* and *three* respectively. The real-world geometrical space is three-dimensional restricted to three mutually perpendicular directions (the Cartesian coordinates x, y and z). The concept of time is included separately as the fourth dimension in the description of evolution processes of three-dimensional real-world systems. However, mathematical models of real-world dynamical systems can have more than three dimensions; the dimensions in this case correspond to the number of degrees of freedom of the system under consideration. The degrees of freedom refer to the independent variables used in the mathematical model, e.g. the flight path of an aeroplane is given by six independent variables, or degrees of freedom, namely, the speed and momentum in the three mutually perpendicular directions at any instant.

1.2.3 Fractal Fluctuations and Statistical Analysis

Most quantitative research involves the use of statistical methods presuming *independence* among data points and Gaussian 'normal' distributions (Andriani and McKelvey 2007). The Gaussian distribution is reliably characterized by its stable mean and finite variance (Greene 2002). Normal distributions place a trivial amount of probability far from the mean and hence the mean is representative of most observations. Even the largest deviations, which are exceptionally rare, are still only about a factor of two from the mean in either direction and are well characterized by quoting a simple standard deviation (Clauset et al. 2007). However, apparently rare real-life catastrophic events such as major earth quakes, stock market crashes, heavy rainfall events, etc., occur more frequently than indicated by the normal curve, i.e. they exhibit a probability distribution with a *fat tail*. Fat tails indicate a power-law pattern and interdependence. The 'tails' of a power-law curve—the regions to either side that correspond to large fluctuations—fall off very slowly in comparison with those of the bell curve (Buchanan 2004). The normal distribution is therefore an inadequate model for extreme departures from the mean.

Fractals are the latest development in statistics. The space-time fluctuation pattern in dynamical systems was shown to have a self-similar or fractal structure. The larger scale fluctuation consists of smaller scale fluctuations identical in shape to the larger scale. An appreciation of the properties of fractals is changing the most basic ways we analyse and interpret data from experiments and is leading to new insights into understanding physical, chemical, biological, psychological, and social systems. Fractal systems extend over many scales and so cannot be characterized by a single characteristic average number (Liebovitch and Scheurle 2000). Further, the self-similar fluctuations imply long-range space-time correlations or interdependence. Therefore, the Gaussian distribution will not be applicable for description of fractal data sets. However, the bell curve still continues to be used for approximate quantitative characterization of data which are now identified as fractal space-time fluctuations (Selvam 2009).

1.2.4 *Golden Mean and Self-similar,* Fractal *Geometrical Structures in Nature*

Animate and inanimate structures in nature exhibit self-similarity in geometrical shape (Stevens 1974; Jean 1994; Freeman 1987, 1990), i.e. parts resemble the whole object in shape. The most fundamental self-similar structure is the forking (bifurcating) structure (Jean 1994) of tree branches, tree roots, river tributaries, branched lightning, etc. The complex branching architecture is a self-similar *fractal* since branching occurs on all scales (sizes) and forms the geometrical shape of the whole object. Self-similar structures incorporate in their geometrical design the *noble numbers*, i.e. numbers, which are functions of the *golden mean* τ and are characterized by fivefold symmetry of the pentagon and dodecahedron. For example, the ratio of the length of the diagonal to the side in a regular pentagon is equal to the *golden mean* τ equal to $(1 + \sqrt{5})/2 \cong 1.618$. The *golden mean* is the most irrational number and is associated with the *Fibonacci* mathematical sequence 1, 1, 2, 3, 5, 8, … where, each term is the sum of the two previous terms and the ratio of each term to the previous term approaches the *golden mean* τ. The *golden mean* τ is the most irrational number in the sense that rational approximations converge very slowly to τ as compared to other irrational numbers. Irrational numbers are numbers such as $\sqrt{2}$, which has an infinite number of non-periodic decimals. Rational approximations such as p/q where p and q are integers are used to represent irrational numbers. The *golden mean* had a special significance in ancient cultures. The significance of the *golden mean* throughout recorded history in science, culture, and religion has been discussed (Hargittai and Pickover 1992; Hargittai 1992). Self-similar spiral structures such as on the shell of the very old mollusk called *Nautilus pompilius* (Jean 1994) incorporate the *golden mean* in their radial growth. *Thompson* described that the *nautilus* followed a pattern originally described by *Rene Descartes* in 1683 as the equiangular spiral and subsequently by *Jacob Bernoulli* as the logarithmic spiral (West 1990a). The commonly found shapes in nature are the helix and the dodecahedron (Stoddart 1988; Muller and Beugholt 1996), which are signatures of self-similarity underlying *Fibonacci* numbers. The association of *noble numbers* with growth of self-similar patterns has been established quantitatively in plant *phyllotaxis* in *botany*. A summary of documented evidence collected over a period of more than 150 years is given below and will help understand the association between *noble numbers* and self-similar patterns in the plant kingdom. *Phyllotaxis* is the study of the arrangement of all plant elements, which originate as primordia on the shoot apex. The botanical elements, which constitute plants, are branches, leaves, petals, stamens, sepals, florets, etc. These plant elements begin their existence as primordia in the neighbourhood of the undifferentiated shoot apex (extremity). Extensive observations in botany show that in more than 90% of plants studied worldwide (Jean 1994; Stewart 1995) primordia emerge as protuberances at locations such that the angle subtended at the apical centre by two successive primordia is equal to the *golden angle* $\varphi = 2\pi\,(1 - 1/\tau)$ corresponding to approximately 137.5°. Theoretical studies show that outside the set of *noble numbers* the structures are not self-similar. The surprisingly precise geometrical placement of plant primordia results in the observed *phyllotactic* patterns, namely, the familiar spiral patterns found in the arrangement of leaves on a stem, in florets of composite flowers, the pattern of scales on pineapple and pine cone, etc. Further, such self-similar patterns ensure identical geometrical design (shape) for all sizes of a single species such as daisy flowers of all sizes. The *phyllotactic* patterns, while pleasing to the eye, also incorporate maximum packing efficiency for fruits and seeds.

1.2.5 Fibonacci *Sequence and Self-similar Structures*

The *Fibonacci* mathematical series was discovered in 1209 by Leonardo of Pisa, known as *Fibonacci* (Kappraff 1992) while computing the total number of adult rabbits in successive months starting with a single adult rabbit pair and assuming that each adult rabbit pair produces one pair of offspring each month and that baby rabbit pairs became adults in one month's time. The growth of rabbit population is shown as a branching network in Fig. 1.3.

The total number of adult rabbit pairs in successive months follows the *Fibonacci* mathematical series. The growth of adult rabbit population as shown in Fig. 1.3 represents a hierarchical ramified network or a self-similar *fractal* network.

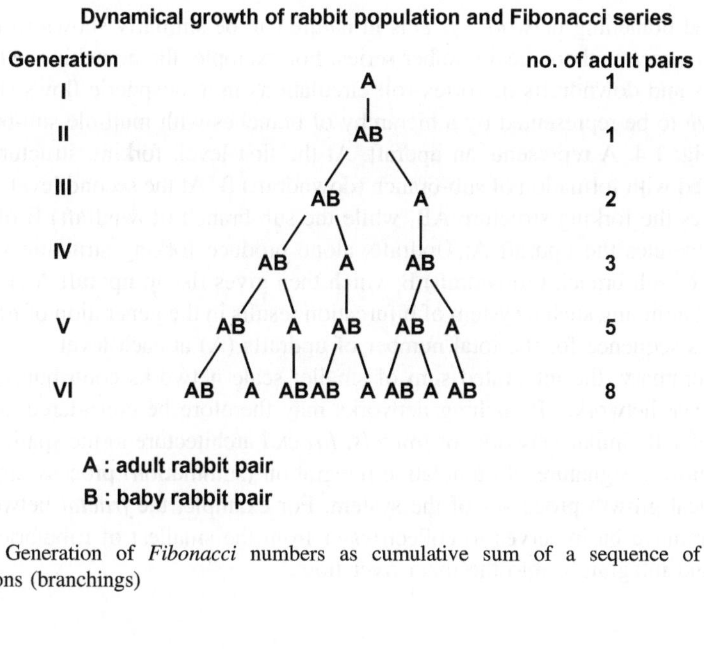

Fig. 1.3 Generation of *Fibonacci* numbers as cumulative sum of a sequence of ordered bifurcations (branchings)

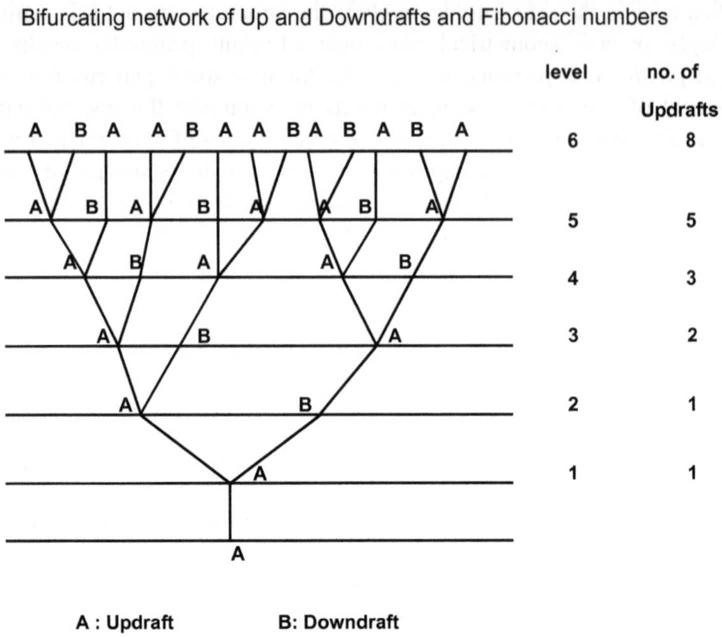

Fig. 1.4 Bifurcating network of updrafts and downdrafts and *Fibonacci* numbers

Ramified branching network systems in nature can be similarly shown to generate the *Fibonacci* mathematical number series. For example, the branching network of updrafts and downdrafts in vortex-roll circulations in atmospheric flows (Fig. 1.4) is shown to be represented by a hierarchy of branches with multiple sub-branches.

In Fig. 1.4, A represents an updraft. At the first level, forking structure AB is generated with formation of sub-branch (downdraft) B. At the second level, A again generates the forking structure AB, while the sub-branch (downdraft) B of level 1 now generates the updraft A. Updrafts alone produce forking structure with formation of sub-branch (downdraft) B which then gives rise to updraft A at the next level. Continuing such a system of bifurcation results in the generation of *Fibonacci* numbers sequence for the total number of updrafts (A) at each level.

In summary, the integrated sum of smaller scale networks contributes to form large-scale networks. Branching networks may therefore be considered as a hierarchy of self-similar networks or *fractals*. *Fractal* architecture to the spatial pattern is therefore a signature of cumulative integration (summation) process inherent to dynamical growth processes of the system. For example, the *fractal* network of a river drainage basin serves to collect water from the smallest of tributaries (branches) and integrate it into the main river flow.

1.2.6 *Fivefold and Spiral Symmetry Associated with Fibonacci Sequence*

The ratio of adjacent elements of the *Fibonacci* sequence approaches the irrational number $\tau = (1 + \sqrt{5})/2$ in the limit. The number, τ, is the solution to the algebraic equation

$$1 + x = x^2$$

As a result τ has the property

$$1 + \tau = \tau^2$$

Therefore, the double geometric sequence $\ldots \frac{1}{\tau^3}, \frac{1}{\tau^2}, \frac{1}{\tau}, 1, \tau, \tau^2, \tau^3, \ldots$ is the *Fibonacci* sequence since it has the property that each term is equal to the sum of the earlier two terms and also the ratio of each term to the earlier term is equal to the *golden mean* τ. It is the only geometric series, which is also a *Fibonacci* sequence (Kappraff 1992). The *Fibonacci* numbers can be represented geometrically in polar coordinates in two dimensions by the equiangular spiral $R_OR_1R_2R_3R_4R_5$ ... drawn with origin O, with lengths of successive radii OR_O, OR_1, OR_2,……… and corresponding spiral segments R_OR_1, R_1R_2, R_2R_3, …… following *Fibonacci* mathematical sequence (Fig. 1.5).

R_OR_1 may be considered to be the tangent at R_O to the circle with centre O and radius OR_O. The angle R_1R_OX equal to α which the tangent makes with the arc XR_O is the crossing angle of the spiral $R_OR_1R_2$ …

The initial radius OR_O equal to R grows to OR_1 equal to $R + dR$ after an angular turning R_OOR_1 equal to $d\theta$. The incremental growth dR is equal to the length R_1X in Fig. 1.5.

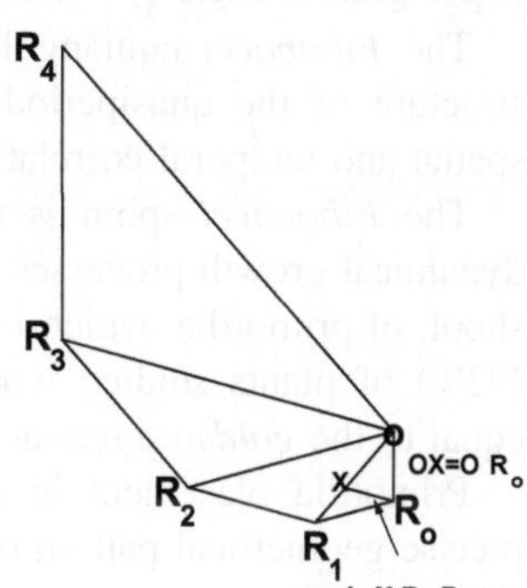

Fig. 1.5 Geometrical representation of *Fibonacci* series in polar coordinates by the *Fibonacci* equiangular spiral

$$\text{Therefore} \quad \tan\alpha = \frac{R_1X}{XR_O}$$

XR_O $Rd\theta$ for the arc XR_O of circle with centre O and radius OR_O equal to R $\tan\alpha$ α in the limit for small values of α

Therefore

$$\frac{dR}{Rd\theta} = \alpha$$
$$\text{or} \quad \frac{dR}{R} = \alpha d\,\theta$$
$$d\ln R = \alpha d\,\theta$$

Integrating for growth of radius from r to R associated with angular turning from 0 to θ,

$$\ln\frac{R}{r} = \alpha\theta$$

or

$$R = re^{\alpha\theta}$$

Geometrical consideration for generation of the *Fibonacci* spiral in three dimensions specify a constant angular turning dθ equal to $\frac{1}{\tau}$ between successive radii and therefore a constant crossing angle, also equal to $\frac{1}{\tau}$. The *Fibonacci* equiangular spiral is then given by the relation

$$R = re^{1/\tau^2}$$

The angle subtended at the centre between two successive radii is therefore equal to the golden angle $\frac{2\pi}{\tau^2}$ or $2\pi\left(1-\frac{1}{\tau}\right)$ since $\frac{1}{\tau^2} = 1 - \frac{1}{\tau}$.

The *Fibonacci* equiangular spiral as shown in Fig. 1.6 has intrinsic internal structure of the quasiperiodic *Penrose tiling pattern* and associated long-range spatial and temporal correlations.

The *Fibonacci* spiral is traced with mathematical precision in nature in the dynamical growth processes of plants as seen in the geometrical placement on the shoot, of primordia, which later develop into the various plant parts. In a majority (92%) of plants studied worldwide, successive primordia always subtend angle equal to the *golden angle* at the apical centre (Jean 1994).

Primordia placement in space and time may therefore be resolved into the precise geometrical pattern of the quasiperiodic *Penrose tiling* pattern.

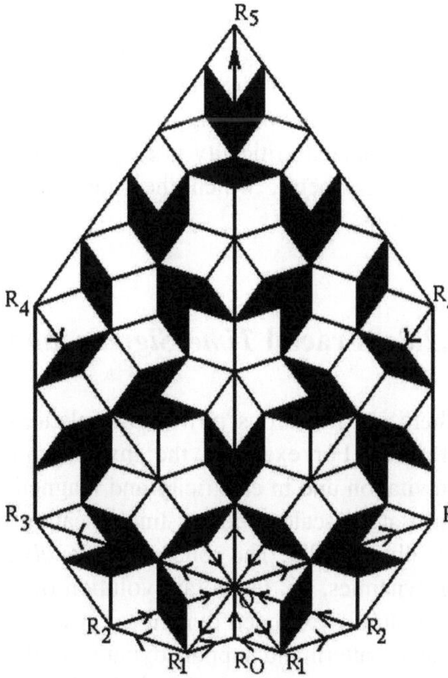

Fig. 1.6 Quasicrystalline structure of the quasiperiodic *Penrose tiling pattern* and *Fibonacci* sequence

1.2.7 Quasicrystalline Structure: The Quasiperiodic Penrose Tiling Pattern

The regular arrangement of plant parts resembles the newly identified (since 1984) *quasicrystalline* order in condensed matter physics (Nelson 1986; Steinhardt 1997; Baake 2002). Traditional (last 100 years) crystallography has defined a crystalline structure as an arrangement of atoms that is periodic in three dimensions. Crystals have lattice structure with identical arrangement of atoms (Von Baeyer 1990; Lord 1991) with space filling cubes or hexagonal prisms. Fivefold symmetry was prohibited in classical crystallography. In 1984, an alloy of aluminium and magnesium was discovered which exhibited the symmetry of an icosahedron with fivefold axis. At the same time Paul Steinhardt of the University of Pennsylvania and his student Dov Levine (Von Baeyer 1990) had quite independently identified similar geometrical structure, now called quasicrystals (Levine and Steinhardt 1984; Mintmire 1996). These developments were based on the important work on the mathematics of tilings done by *Roger Penrose* and others beginning in the 1970s. Penrose (1974, 1979) discovered a non-periodic tiling of the plane, using two types of tiles, which is a quasiperiodic crystal with pentagonal symmetry (DiVincenzo 1989). It is generally accepted that a quasicrystal can be understood as a systematic (but not periodic) filling of space by unit cells of more than one kind. Such extended structures in space can be orderly and systematic without being periodic. *Penrose tiling* patterns (Fig. 1.6) are two-dimensional quasicrystals.

The geometric pattern is self-similar and exhibits long-range correlations and is quasiperiodic. Selvam (1990) has shown that turbulent fluid flows can be resolved into the quasiperiodic *Penrose tiling* pattern with *fractal* self-similar geometry to spatial pattern and long-range temporal correlations for temporal fluctuations. Self-organized criticality is exhibited as the *Penrose tiling* pattern for self-similar spatial geometry, which then incorporates temporal correlations for dynamical processes.

1.2.8 Fractal *Time Signals, and Power Laws*

There are numerous power-law relations in science that have the self-similarity property. For example, the inverse square-law force, which is fundamental in gravitation and in electricity and magnetism, has no intrinsic scale. It has the same form at all scales under a linear scaling transformation (Deering and West 1992; Wienberg 1993). The concept of *fractals* may be used for modelling certain aspects of dynamics, i.e. temporal evolution of spatially extended dynamical systems.

Spatially extended dynamical systems in nature exhibit *fractal* geometry to the spatial pattern and support dynamical processes on all timescales, for example, the *fractal* geometry to the global cloud-cover pattern is associated with fluctuations of meteorological parameters on all timescales from seconds to years. The temporal fluctuations exhibit structure over multiple orders of temporal magnitude in the same way that *fractal* forms exhibit details over several orders of spatial magnitude. The power spectra of such broadband fluctuations exhibit inverse power law of form $1/f^{\alpha}$ where f is the frequency and α the exponent. In general, α decreases with f and approaches 1 for low frequencies. Self-similar variations on different timescales will produce a frequency spectrum having an inverse ($1/f$) power-law distribution or $1/f$-like distribution and imply long-range temporal correlations signifying persistence or 'memory'. The frequency range over which α is constant therefore exhibits self-similarity or scale invariance in temporal fluctuations, i.e. the fluctuations are fractals in time. The intensity or variance of longer and shorter period fluctuations are mutually related by a scale factor alone independent of the nature of dynamical processes. The fluctuations exhibit long-range temporal correlations. Also, temporal fluctuations exhibit multifractal structure since α varies for different ranges of frequency f.

The phenomenon of $1/f$-noise spectrum first introduced by Van Der Ziel in 1950 (West and Shlesinger 1989) is ubiquitous to dynamical systems in nature and has a long history of more than 40 years of observational documentation in all fields of science and other areas (Bak 1988; West 2004; Milotti 2002; Li 2007). The multidisciplinary nature of investigations will help gain new insights and develop mathematical and statistical techniques and analytical tools for understanding and quantifying the physics of the observed long-range correlations in dynamical systems in nature. The physics of dynamical systems therefore comes under the broad category of general systems theory. The subunits of the system function as a unified whole two-way communication and control network with global (system level) control/response to local functions/stimuli, thereby possessing the criteria for a robust system (Csete and Doyle 2002; Kitano 2002, 2004). Kitano (2002) makes the point that robustness is a property of an entire system; it may be that no individual component or process within a system would be robust, but the system-wide architecture still provides robust behaviour. This presents a challenge for analysis, since elucidating such behaviours can be counterintuitive and computationally demanding.

Power-law behaviour has been documented in the functioning of physiological systems (Sun and Charef 1990; Suki et al. 1994). Hurst (1951) and Hurst et al. (1965) had shown for river flows (Schepers et al. 1992) that for a wide variety of data sets the degree of 'memory' over time spans of up to a millennium could be characterized by a power-law relationship (Bassingthwaighte and Beyer 1991). Long-range spatial correlations have been identified at the level of the DNA (Maddox 1992; Peng et al. 1992). Long-range correlations over time and space have also been investigated by Mandelbrot and Wallis (1969) for geophysical records and more recently by Tang and Bak (1988), and Bak et al. (1987, 1988), for $1/f$ noise in dynamical systems. Andriani and McKelvey (2007) have given exhaustive references to earliest known work on power-law relationships. A power-law world is dominated by extreme events ignored in a Gaussian world. In fact, the fat tails of power-law distributions make large extreme events orders of magnitude more likely. Theories explaining power laws are also scale-free. This is to say, the same explanation (theory) applies at all levels of analysis (Andriani and McKelvey 2007).

The $1/f$ power law would seem to be natural and white noise (flat distribution) would be the subject of involved investigation (West and Shlesinger 1989). Recent studies have identified turbulent cascades in foreign exchange markets (Ghashghaie et al. 1996) and power laws governing epidemics have been reported (Rhodes and Anderson 1996). Universality gives a new understanding of how apparently very different things can act in the same way (Buchanan 1997).

A major feature of this correlation is that the amplitude of short-term- and long-term fluctuations are related to each other by the scale factor alone independent of details of growth mechanisms from smaller to larger scale. The macroscopic pattern, comprised of a multitude of sub-units, functions as a unified whole independent of details of dynamical processes governing its individual sub-units (Mantegna and Stanley 1995). Such a concept that physical systems, which consist of a large number of interacting sub-units, obey universal laws that are independent of the microscopic details is now acknowledged as a breakthrough in statistical physics. The variability of individual elements in a system act cooperatively to establish regularity and stability in the system as a whole (West and Shlesinger 1989). Scale invariance implies, knowledge of the properties of a model system at short times or short length scales can be used to predict the behaviour of a real system at large times and large length scales (Stanley 1996a).

The *fractal* dimension D of a temporal *fractal* can be computed using recently developed algorithms. Since time series of a single variable such as temperature in atmospheric flows may reflect the cumulative effect of the multitude of factors governing flow dynamics, the *fractal* dimension may indicate the number of parameters controlling the evolution dynamics. However, a knowledge of D alone does not help identification of the parameters or their exact role in the dynamical growth processes. Also, limitations in data length and computational algorithms preclude exact determination of D (Lorenz 1991). It has not been possible to formulate governing equations based on a knowledge of D for prediction purposes.

1.2.9 Self-organized Criticality: Space-Time Fractals

Till very recently (1987), fractal geometry to the spatial pattern and fractal fluctuations in time of dynamical processes of the same extended dynamical system were treated as two disparate multidisciplinary fields of research (Bak and Chen 1989). The long-range spatiotemporal correlations underlying spatial and temporal power-law behaviour of dynamical systems was identified as a unified manifestation of self-organized criticality (SOC) in 1987 (Bak et al. 1987, 1988; Bak and Chen 1989; Stanley 1995).

The unifying concept of self-organized criticality underlying fractals, self-similar scaling, broadband frequency spectra and inverse power law distribution offer new and powerful means of describing certain basic aspects of spatial form and dynamical processes in a dynamical system. The systems in which self-organized criticality is observed range from the physical to the biological to the social. The physical mechanism underlying the observed self-organized criticality is not yet identified. However, the long-range spatial and temporal correlations underlying dynamical evolution implies predictability in space and time of the pattern of evolution of the dynamical system, for example, atmospheric flows.

The relation between spatial and temporal power-law behaviour was recognized much earlier in condensed matter physics where long-range spatiotemporal correlations appear spontaneously at the critical point for continuous phase transitions. The amplitudes of large and small-scale fluctuation are obtained from the same mathematical function using appropriate scale factor, i.e. ratio of the scale lengths. This property of self-similarity is often called a renormalization group relation in physics (Wilson 1979; West 1990a; Peitgen et al. 1992) in the area of continuous phase transitions at critical points (Weinberg 1993; Back et al. 1995). When a system is poised at a critical point between two macroscopic phases, e.g. vapour to liquid, it exhibits dynamical structures on all available spatial scales, even though the underlying microscopic interactions tend to have a characteristic length scale (Back et al. 1995). But, in order to arrive at the critical point, one has to fine-tune an external control parameter, such as temperature, pressure or magnetic field, in contrast to the phenomena described above which occur universally without any fine-tuning. The explanation is that open extended dissipative dynamical systems i.e. systems not in thermodynamic equilibrium may go automatically to the critical state as long as they are driven slowly: the critical state is self-organized (Tang and Bak 1988; Bak and Chen 1989, 1991).

Fluctuations in time of atmospheric flows, as recorded by meteorological parameters such as pressure, temperature, wind speed, etc., exhibit self-similar fluctuations in time, namely, a zigzag pattern of increase (decrease) followed by a decrease (increase) on timescales from seconds to years. Such jagged pattern for atmospheric variability (temporal) resembles the self-similar coastline structure. Long-range correlations in space and time, namely self-similar (fractal) fluctuations in space and time implies that the magnitude of the fluctuation (spatial or temporal) at any scale is related to other scales (larger and smaller) by a single parameter, namely, the scale factor which is a dimensionless number. Therefore, dynamical laws which govern the space-time fluctuations of smallest scale (turbulence, millimetres/seconds) fluctuations in space-time also apply for the largest scale (planetary, thousands of kilometres/years) in atmospheric flows throughout the globe. The cooperative existence of fluctuations of all scales gives rise to self-similar (coherent) space-time structures. The formation of such coherent structures which function as a unified whole has a special significance in the field of *Biology*, in the functioning of living systems. Pattern formation (Dangelmayr 1987), i.e. morphogenesis, forms an integral part of *Life Sciences* and the vast amount of knowledge gathered in this field can beneficially be applied to other fields of science since self-similar space-time patterns are generic to nature, in particular, weather and climate patterns in meteorology. The multidisciplinary approach to the study of self-organized criticality will result in immense benefit to the scientific community as a whole in terms of transfer of new insights and techniques from one field to another.

1.2.10 Turbulent (Chaotic) Fluctuations and Self-similar Structure Formation

The first phenomenological treatments of morphogenesis were built for fluid dynamics through the mathematical modelling of instabilities as those named after Faraday, Rayleigh and Bernard, Rayleigh and Taylor, Kelvin and Helmholtz, etc., (Arecchi 1995). Biological auto-organization and pattern formation have been studied over the past 40 years as non-equilibrium thermodynamic phenomena (Turing 1952). Biological systems exhibit high degree of cooperation in the form of long-range communication. The concept of cooperative existence of fluctuations in the organization of coherent structures have been identified as self-organized cooperative phenomena (Prigogine 1980). The study of the spontaneous, i.e. self-organized formation of structures in systems far from thermal equilibrium in open systems belongs to the multidisciplinary field of *synergetics* (Haken 1980, 1989). Plant kingdom exhibits examples of the most striking self-similar geometrical patterns (Jean 1994) signifying *self-organized criticality* in the spatial structure formation.

Formation of structure begins by aggregation of molecules in a turbulent fluid (gas or liquid) medium. Turbulent fluctuations are therefore not dissipative, but serve to assemble and form coherent structures (Nicolis and Prigogine 1977; Prigogine and Stengers 1988; Insinnia 1992), for example, the formation of clouds in turbulent atmospheric flows (Tennekes 1973). Traditionally, turbulence is considered dissipative and disorganized. Yet, coherent (organized) vortex-roll circulations (vortices) are ubiquitous to turbulent fluid flows (Levich 1987; Frisch and Orszag 1990). The exact physical mechanism for the formation and maintenance of coherent structures, namely vortices or large eddy circulations in turbulent fluid flows is not yet identified. The most intense weather systems such as hurricanes have vividly spiralling cloud formation while the destructive tornado has spiralling (vortex) air flow in narrow funnel-like protuberances which reach down to earth and create devastating damage.

Recent studies show that clouds of all sizes (Tessier et al. 1993) are self-similar in shape, which is consistent with commonly visualized shape of clouds as billows upon billows. Incidentally, it may be mentioned that cumulus clouds bear a close resemblance to cauliflowers. Meteorological textbooks commonly describe the cumulus clouds to have cauliflower-like structure. In the midst of turbulence in air flows, clouds retain their identity in shape and the most astonishing of all is the formation of ice crystals with exquisitely symmetrical structure.

Nature abounds in symmetrical structures from the macro- to the microscopic scales (Tarasov 1986). Perfect order appears to underlie apparent chaos in turbulent flows. Turbulence, namely, seemingly random fluctuations of all scales, therefore, plays a key role in the formation of self-similar coherent structures in the atmosphere. Such a concept is contrary to the traditional view that turbulence is dissipative, i.e. ordered growth of coherent form is not possible in turbulent flows. Selvam (1990), Selvam and Fadnavis (1998), Selvam et al. (1992), Joshi and Selvam (1999) have shown that turbulent fluctuations self-organize to form self-similar structures in fluid flows.

Ramified branching networks serve to connect and assist in the functioning as a unified whole of self-similar *fractal* objects. A *fractal* object can be resolved into smaller interconnected component parts, which resemble the whole in shape. The self-similar architecture for *fractal* objects serves for collection and distribution of information/energy between the largest and smallest scales. For example, the river system collects water from tributaries, the lung architecture enables efficiency of oxygen absorption from air in the alveoli (the smallest branching structure). Jean (1994) has emphasized the functional importance of ramified structures underlying self-similar *fractals* and gives reference to earlier studies, which show that such branching structures can be organized into hierarchies, which incorporate the *Fibonacci* mathematical sequence.

1.2.11 *Self-similarity: A Signature of Identical Iterative Growth Process*

Self-similarity underlies all growth processes in nature. Jean (1994) has emphasized the self-similar geometry of botanical elements. Self-similar structures are generated by iteration (repetition) of simple rules for growth processes on all scales of space and time. Such iterative processes are simulated mathematically by numerical computations such as $X_{n+1} = F\ (X_n)$ where X_{n+1}, the value of the variable X at $(n + 1)$th computational step is a function F of its earlier value X_n. Mathematical models of real-world dynamical systems are basically such iterative computational schemes implemented on finite precision digital computers. Computer precision imposes a limit (finite precision) on the accuracy (number of decimals) for numerical representation of X. Since X is a real number (infinite number of decimals) finite precision introduces round-off error in iterative computations from the first stage of computation. The model iterative dynamical system therefore incorporates round-off error growth. Computed growth patterns exhibit self-similar *fractal* structure which incorporates the *golden mean* (Stewart 1992a). The new science of *nonlinear dynamics and chaos* seeks to understand the physics of such self-similar patterns in computed and real-world dynamical systems.

1.3 Fractals and Self-organized Criticality in Meteorology and Atmospheric Physics

Fluid flows such as river flows, atmospheric flows, etc., are characterized by turbulence, namely, seemingly random fluctuations on all space and timescales. Traditionally, turbulence is considered dissipative and disorganized. Yet, coherent (organized) vortex-roll circulations (vortices) are ubiquitous to turbulent fluid flows (Tennekes 1973; Levich 1987; Frisch and Orszag 1990). The exact physical mechanism for the formation and maintenance of coherent structures, namely vortices or large eddy circulations in turbulent fluid flows is not yet identified. Paradoxically, the more severe the turbulence (disorder), the more vivid the organization of coherent spiralling structures, e.g. the spiral cloud bands, and airflow circulations in the destructive hurricanes and tornadoes. Lovejoy and Schertzer (2010) have shown conclusively that the seemingly irregular fluctuations of meteorological parameters are self-similar *fractals*, the power spectra exhibiting inverse power-law form $f^{-\alpha}$ where f is the frequency and the exponent α is different for different scale ranges. The fluctuations exhibit scale invariance or self-similarity for scale ranges with constant scale factor α.

Nonlinear dynamical systems in nature such as atmospheric flows exhibit complex spatial patterns, e.g. cloud geometry, that lack a characteristic (single) length scale concomitant with temporal fluctuations that lack a single timescale. Objects in nature are in general *multifractals*, i.e. the *fractal dimension* is different for different scale ranges. The concept of *fractal dimension* introduced by Mandelbrot (1977) provides a powerful tool for quantitative description of nonlinear fluctuations in real-world and computed dynamical systems. Self-similar fractal structures are also found in finite precision computer realization of nonlinear mathematical models of dynamical systems. Self-similarity implies long–range spatiotemporal correlations. Lorenz (1963) was the first to identify the sensitive dependence on initial conditions characteristic of such self-similar structures in finite precision computer realization of a simple model of atmospheric flows. Sensitive dependence on initial conditions precludes exact prediction and therefore named *deterministic chaos* (Gleick 1987) since deterministic model equations give chaotic numerical solutions. Numerical models, even with only a few degrees of freedom resemble real-world dynamical systems in generating irregular (complex) fluctuations. The concept of *fractals*, i.e. self-similar fluctuations implies long-range correlations in space and time. Long-range spatiotemporal correlations are ubiquitous to dynamical systems in nature and are identified as signatures of self-organized criticality (Bak et al. 1988; Bak and Chen 1991; Stanley 1995). The fractal structure of atmospheric flows in space and time has been identified and discussed in detail by Lovejoy and his group (Lovejoy and Schertzer 1986a, b; Schertzer and Lovejoy 1991, 1994; Tessier et al. 1993 and the references therein). Fractals and multifractals characterize fluid turbulence (Sreenivasan 1991), atmospheric flows being a representative example of fluid turbulence. Atmospheric flows therefore exhibit *self-organized criticality*. Standard meteorological theory cannot explain satisfactorily (Tessier et al. 1993) the observed self-similar structures to spatiotemporal pattern of atmospheric flows ranging from the turbulence (millimetres/seconds) to climatological (kilometres/years) scales.

The cooperative existence of fluctuations ranging in size-duration from a few millimetres/seconds (turbulence scale) to thousands of kilometres/years (planetary scale) result in the observed long-range spatiotemporal correlations, namely, fractal geometry to the global cloud-cover pattern concomitant with inverse power-law form for power spectra of temporal fluctuations documented by Lovejoy and Schertzer (1986a, b) and Tessier et al. (1993). Long-range spatiotemporal correlations are ubiquitous to real-world dynamical systems and are recently identified as signatures of self-organized criticality (Bak et al. 1988). The physics of self-organized criticality is not yet identified. It is important to quantify the total pattern of fluctuations in atmospheric flows for predictability studies. Traditional numerical weather prediction models based on Newtonian continuum dynamics are nonlinear and require numerical solutions incorporating numerical integration schemes which are basically iterative computations. Finite precision computer realizations of such nonlinear models are sensitively dependent on initial conditions, now identified as deterministic chaos (Gleick 1987) resulting in unrealistic solutions. The physics of deterministic chaos is not yet identified. Selvam (1993) has shown that round-off error approximately doubles on an average for each step of finite precision numerical iteration. Such round-off error doubling results in unrealistic solutions for numerical weather prediction and climate models which incorporate long-term numerical integration schemes with thousands of such iterations. Realistic modelling of atmospheric flows therefore requires alternative concepts for fluid flows and robust computational techniques which do not require round-off error prone calculus-based long-term numerical integration schemes.

1.3.1 Observed Structure of Atmospheric Flows and Signatures of Deterministic Chaos

Recent advances in remote sensing and in situ measurement techniques have atmospheric scientists to document the following new observational characteristics of turbulent shear flows in the planetary atmospheric boundary layer (ABL) where weather activity occurs. The ABL extends to about 10 km above the surface of the earth.

(i) The atmospheric flow consists of a full continuum of fluctuations ranging in size from the turbulence scale of a few millimetres to the planetary scale of thousands of kilometres.

(ii) The atmospheric eddy energy spectrum follows an inverse power law of form $f^{-\alpha}$ where f is the frequency and α the exponent. The exponential power-law form for the eddy energy spectrum indicates self-similarity and scale invariance. The exponent α is found to be equal to 1.8 for both meteorological (time period in days) and climatological (time period in years) scales, which indicates a close coupling between the two scales (Dewan and Good 1986; Van Zandt 1982; Lovejoy and Schertzer 1986a, b; Fritts and Chou 1987; Canavero and Einaudi 1987; Dewan et al. 1988; Tsuda 1989).

(iii) Satellite cloud-cover photographs give evidence for the existence of helical vortex-roll circulations (or large eddies) in the ABL as indicated by the organization of clouds in rows and (or) streets, meso–scale (up to 100 km) cloud clusters (MCC), and spiral bands in synoptic scale weather systems (Eymard 1985).

(iv) The structure of atmospheric flows is invariably helical (curved) as manifested in the visible cloud patterns of weather systems, e.g. all basic mesoscale structures such as medium scale tornado generating storms, squall lines, hurricanes, etc. (Levich 1987), and in particular the supercell storm (Lilly 1986).

(v) Atmospheric flows give an implicit indication of the upscale transfer of a certain amount of energy inserted at much smaller scales, thereby generating the observed helical fluctuations (Levich 1987; Lilly 1989).

(vi) The global cloud-cover pattern exhibits self-similar fractal geometrical structure and is consistent with the observed scale invariance of the atmospheric eddy energy spectrum (Lovejoy and Schertzer 1986a, b; Schertzer and Lovejoy 1989) (see characteristic (*ii*) above).

Atmospheric weather systems exist as coherent structures consisting of discrete cloud cells forming patterns of rows and (or) streets, mesoscale cloud clusters (MCC), and spiral bands. These patterns maintain their identity for the duration of their appreciable lifetimes in the apparently dissipative turbulent shear flows of the ABL (Tennekes 1973). The existence of coherent structures (seemingly systematic motion) in turbulent flows in general, has been well established during the last 30–40 years of research into turbulence. However, it is still debated whether these structures are the consequences of some kind of instabilities (such as shear or centrifugal instabilities), or whether they are manifestations of some intrinsic universal properties of any turbulent flow (Levich 1987).

Lovejoy and Schertzer (1986a, b) have provided conclusive evidence for the signature of deterministic chaos in atmospheric flows, namely the fractal geometry of global cloud-cover pattern and the inverse power law form $f^{-\alpha}$ where f is the frequency and α the exponent for the atmospheric eddy energy spectrum. Atmospheric teleconnections, such as the El Nino and (or) Southern Oscillation (ENSO) cycles in weather patterns, that are responsible for devastating changes in normal global weather regimes (Trenberth et al. 1988; Kushnir and Wallace 1989; Lau et al. 1989) are also manifestations of long-range correlations in regional weather activity.

Meteorologists have documented in detail the nonlinear variability of atmospheric flows, in particular the interannual variability, i.e. the year to year fluctuations in weather patterns. The interannual variability of atmospheric flows is nonlinear and exhibits fluctuations on all scales ranging up to the length of data period (time) investigated. The broadband spectrum of atmospheric interannual variability has embedded dominant quasiperiodicities such as the quasibiennial oscillation (QBO) and the ENSO (*El Nino/Southern Oscillation*) cycle of 3–7 years which are identified as major contributors to local climate variability, in particular, the monsoons which influence agriculture dependent world economies. ENSO is an irregular (3–7 years), self-sustaining cycle of alternating warm and cool water episodes in the Pacific Ocean. Also called *El Nino—La Nina*, *La Nina* refers to the cool part of the weather cycle while *El Nino* is associated with a reversal of global climatic regimes resulting in anomalous floods and droughts throughout the globe. It is of importance to quantify the total pattern of fluctuations for predictability studies.

1.3.2 Limitations of Conventional Atmospheric Boundary Layer (ABL) Models

Presently available models for ABL turbulent flows are incapable of identifying the coherent helical structural form intrinsic to turbulence. Also, the models do not give realistic simulations of the space-time averages for the thermodynamic parameters and the fluxes of buoyant energy, mass, and momentum because of the following inherent limitations:

(i) The physics of the observed coherent helical geometric structure inherent in turbulent flows is not yet identified, and therefore the structural form of turbulent flows cannot be modelled.
(ii) By convention, the Newtonian continuum dynamics of the atmospheric flows are simulated by the Navier–Stokes (NS) equations, which are inherently nonlinear, and being sensitive to initial conditions, give chaotic solutions characteristic of deterministic chaos.
(iii) The governing equations do not incorporate the mutual coexistence and interaction of the full spectrum of atmospheric fluctuations that form an integral part of atmospheric flows (Lorenz 1986; Fritts and Chou 1987; Shaffee and Shaffee 1987; Shepherd 1987).
(iv) The limitations of available computer capacity necessitate severe truncations of the governing equations, thereby generating errors of approximations.
(v) The above-mentioned uncertainties are further magnified exponentially with time by computer round-off errors and result in unrealistic solutions (Beck and Roepstorff 1987; Mccauley 1988). Recent exhaustive studies by Weil (1985) and others also indicate that existing numerical models of atmospheric boundary layer flows require major revisions to incorporate an understanding of turbulence and diffusion in boundary layer flows. Recently, there has been growing conviction that current numerical weather prediction models are inadequate for accurate forecasts (Grebogi et al. 1988; Lighthill 1986; Mason 1986; Reinhold 1987; Kerr 1989; Lovejoy and Schertzer 2010). Numerical modelling of atmospheric flows, diffusion, and cloud growth therefore require alternative concepts and computational techniques.

1.3.3 Traditional Numerical Weather Prediction, Deterministic Chaos and Predictability

Standard models for turbulent fluid flows in meteorological theory cannot explain satisfactorily the observed *multifractal* (space-time) structures in atmospheric flows (Tessier et al. 1993, 1996). Traditionally, meteorological theory is based on the following concepts. The turbulent atmospheric flows are governed by the mutual interaction of a large number of factors, i.e. variables such as pressure, temperature, moisture content, wind speed, etc. Historically, Richardson (1965), in 1922, formulated quantitative computational method for weather prediction as follows. The prediction of future flow pattern is based on mathematical equations for the rate of change dX/dt of component variable X with time t. The rate of change with time dX/dt of any variable X is generally a nonlinear function of all the other interacting variables and therefore analytical solution for X is not available. The evaluation of any variable X with time is then computed numerically from the iterative equation

$$X_{n+1} = X_n + \left(\frac{dX}{dt}\right)_n dt,$$

where the subscript n denotes the time step and the rate of change dX/dt is assumed to be continuous for small changes in time dt, an assumption based on *Newtonian* continuum dynamics. The successive values of X are then computed iteratively, a process known as numerical integration. The prediction equation for the variable X has intrinsic error feedback loop since the value of X at each step is a function of its earlier value in such numerical integration computational techniques. The fundamental (basic) error in numerical computations is the round-off error of finite precision computations. Blank (1994) mentions that when solving differential or other dynamical systems on a computer, the effects of finiteness (round-off) can sometimes be very drastic. When we work with fixed precision system, not all real numbers are even representable and arithmetic does not have the properties that we are used to (Corless et al. 1990).

Lorenz (1989) has discussed chaotic behaviour when continuum equations are solved numerically as difference equations. Climate modelling concepts has come under criticism lately since uncertainty in input parameter values can give drastically different results (Kerr 1994). Selvam (1993) has shown that round-off error approximately doubles on an average at each step of iteration. Such error doubling at each step in numerical integration will result in the round-off error propagating into the mainstream (digits place and above) computation within 50 iterations using single precision (seventh decimal place accuracy) digital computers. In addition, any uncertainties in specifying the initial value of the variable X will also grow exponentially with time and give unrealistic solutions. Numerical solutions are therefore sensitively dependent on initial conditions. Deterministic governing equations, namely evolution equations which are precisely defined and mathematically formulated give chaotic solutions because of sensitive dependence on initial conditions. Finite precision computer realizations of nonlinear mathematical models of dynamical systems therefore exhibit deterministic chaos. Computed model solutions are therefore mere mathematical artifacts of the universal process of round-off error growth in iterative computations (Selvam 1993) and the computed domain is the successive cumulative integration of round-off error domains analogous to the formation of large eddy domains as envelopes enclosing turbulent eddy fluctuation domains such as in atmospheric flows (Selvam 1990, 2005, 2007, 2009, 2011, 2013).

Computed solutions, therefore qualitatively resemble real-world dynamical systems such as atmospheric flows with manifestation of *self-organized criticality*. *Self-organized criticality*, i.e. long-range spatiotemporal correlations, originates with the primary perturbation domains corresponding respectively to round-off error and dominant turbulent eddy fluctuations in model and real-world dynamical systems. Computed solutions, therefore, are not true solutions. The vast body of the literature investigating chaotic trajectories in recent years (since 1980) documents only the round-off error structure in finite precision computations. Stewart (1992b) mentions that in the absence of analytical (true) solutions the fidelity of computed solutions is questionable.

Hacker et al. (2005) discuss predictability of dynamical systems in general and atmospheric flows in particular as follows. The study of predictability is multi-faceted and appears in diverse fields. For purposes of discussion following the lead of Tribbia and Baumhefner (1988) Hacker et al. (2005) adopted the definition of predictability proposed by Thompson (1957) which is 'the extent to which it is possible to predict (the atmosphere) with a theoretically complete knowledge of the physical laws governing it'. More precisely, Hacker et al. (2005) interpreted this as the state-dependent rate of divergence of trajectories in phase space given complete knowledge of system dynamics. Therefore, predictability is intrinsic to a system, and the atmosphere (most likely) has predictability properties distinct from those of any model. Similar statements can be made about biological and all other dynamical systems. We can exactly describe and solve for the evolution of some simple systems analytically, but we are faced with the frustrating reality that we cannot precisely know the predictability of more complex systems. Thus, much of our science is the pursuit of an unknowable goal (Hacker et al. 2005).

Zupanski and Navon (2006) state that uncertainty estimation is becoming an important new research discipline, crosscutting many scientific areas. The mathematical concept of uncertainty estimation is based on probability theory and statistics, estimation theory, information theory, and control theory. Theoretical aspects of uncertainty estimation are generally well understood for linear models (operators) and Gaussian distribution. In the geosciences, however, nonlinear models are typically used; thus, the Gaussian probability assumption may not be the best option. In addition, models of geosciences systems are typically high-dimensional, with state variable dimensions of the order of 10^6–10^7. At the same time, the mathematical concept of uncertainty estimation, algorithmically defined by smoothing and/or filtering, is relatively simple, and a common mathematical framework can be applied across disciplines. These facts create a challenging problem for uncertainty estimation, requiring new scientific developments and cross-disciplinary efforts.

Historically, deterministic chaos, the origin of uncertainty in computed solutions was identified nearly a century ago by *Poincare* in his study of the three body problem (Poincare 1892). Lack of high-speed computational machines precluded exhaustive studies of nonlinear behaviour and approximate linearized solutions of nonlinear systems alone were studied. With the advent of electronic digital computers in late 1950s, Lorenz (1963) identified deterministic chaos in a simple model of atmospheric flows. *Lorenz*'s result captured the attention of scientists in all branches of science since a simple set of equations exhibits chaotic behaviour similar to the complex, irregular (unpredictable) fluctuations exhibited by real-world dynamical systems. Till then it was believed that complex behaviour results from complexity in the governing parameters and the mathematical formulations. *Lorenz*'s model demonstrated that simple models can demonstrate complex behaviour of real-world dynamical systems.

The computed trajectory is plotted graphically in phase space of dimension *m* where *m* is the number of variables representing the dynamical system. For example, a particle in motion can be represented completely at any instant by its position and momenta in the *x, y and z* directions, i.e. six-dimensional phase space. The line joining the successive points in time gives the trajectory of the particle in phase space. The trajectory traces the *strange attractor*, so named because of its strange convoluted shape being the final destination of all trajectories in the phase space. Two trajectories, initially close together diverge exponentially with time though still within the *strange attractor* domain, thereby exhibiting sensitive dependence on initial conditions or *deterministic chaos*. The *strange attractor* exhibits self-similar *fractal* geometry similar to the space-time *fractal* structure or *self-organized criticality* exhibited by real-world dynamical systems. Selvam (1993) has shown that the *strange attractor* has the quasicrystalline structure of the quasiperiodic *Penrose* tiling pattern. There is a very close similarly between the geometrical patterns generated during iterative computations and those found in nature (Jurgen et al. 1990; Stewart 1992a). Iterative computations generate patterns strongly reminiscent of plant forms and clearly these curious configurations show that the rules responsible for the construction of elaborate living tissue structures could be absurdly simple (Dewdney 1986). In summary, self-similar space-time structures or *self-organized criticality* is ubiquitous to dynamical systems in nature and also to mathematical models of dynamical systems which incorporate finite precision iterative computations with resultant feedback and magnification of round-off error primarily, in addition to initial errors. Iterative computations result in the cumulative addition (integration) of the progressively increasing round-off error. Persistent perturbations, though small in magnitude are therefore capable of generating complex space-time structures with *fractal* self-similar geometry because of feedback with amplification.

1.3.4 Current Techniques in Numerical Weather Prediction (NWP): Major Drawbacks

Present-day weather/climate predictions are probabilistic ensemble forecasts. Atmospheric evolution is chaotic, i.e. sensitive to initial-condition uncertainty. However, with modern-day supercomputers, we can run weather forecast models many times from very slightly different initial conditions, consistent with the uncertainties to estimate the effect of this initial-condition uncertainty. The resulting forecasts can be combined to produce a forecast probability distribution and is basically an ensemble weather forecast (Palmer 2005).

Roebber and Tsonis (2005) describe the ensemble forecasting method and its drawbacks as follows. Ensemble prediction has become an indispensable tool in weather forecasting. One of the issues in ensemble prediction is that, regardless of the method, the prediction error does not map well to the underlying physics (i.e.

error estimates do not project strongly onto physical structures). The fundamental problem of weather forecasting is to identify the range of possible meteorological scenarios that might evolve from a given initial state, and determine whether multiple solutions have high probability (low confidence in an individual solution) or if a single evolution is the most likely (high confidence). This probabilistic view is necessitated by the complexity of atmospheric dynamics (e.g. Lorenz 1963; see a general review in Kalnay 2003). Specifically, limits to deterministic predictability originate from two general sources: model error and initial condition error. Model error arises because of imperfections in our characterizations of the laws of nature, arising either through parameterization of complex and/or poorly understood physics (such as boundary layer and cloud microphysical processes) or an inability to resolve atmospheric processes smaller than a certain threshold (e.g. atmospheric convection with a 10 km gridpoint model), with resultant upscale error growth. Initial condition error arises because of the finite spatial availability of observed data (including some variables that are not observed at all), missing data, inaccuracies in the data, and imperfect analysis techniques. All of these errors, even with a perfect model, will grow nonlinearly over time, eventually swamping the forecast signal (Lorenz 1963, 1965, 1969). The rate of this error growth and hence the lead time at which predictability is lost depends on the stability of the evolving flow (Lorenz 1965), which in addition is affected by factors such as the large-scale flow pattern, season, and geographical domain (Lorenz 1984, 1990). Ensemble forecast systems have been developed as a means to quantify forecast uncertainty, using a variety of methods to simulate analysis and model uncertainties (Roebber and Tsonis 2005).

Lovejoy and Schertzer (2010) have summarized the current status of NWP as follows. Twenty years ago the goal of weather forecasting was to determine the (supposedly unique) state of the atmosphere at some time in the future, whereas today, ensemble forecasting systems have instead the goal of determining the possible states of tomorrow's weather including their probabilities of occurrence. This new goal therefore corresponds to a transition from deterministic to stochastic forecasts. Today's ensemble forecasting systems therefore require knowledge of the underlying stochastic structure of the deterministic equations.

The current ensemble forecasting technique is essentially a stochastic–deterministic hybrid which is indirect and problematic on several counts. The main difficulties are (i) that it is based on a deterministic framework for the initial objective analysis—which uses statistics only to describe measurement errors—and not the fields themselves—and (ii) which assumes that the fields evolve according to deterministic nonlinear partial differential equations. While deterministic assumptions may be appropriate for descriptions and models at the dissipation scale, stochastic ones are more appropriate at lower space-time resolution (if only because an infinite number of different dissipation scale fields give rise to the same low resolution analysis fields).

Lovejoy and Schertzer (2010) have discussed the issue of distinguishing natural from anthropogenic variability and the problem of outliers as follows. Conclusions about anthropogenic influences on the atmosphere can only be drawn with respect to the null hypothesis, i.e. one requires a theory of the natural variability, including knowledge of the probabilities of the extremes at various resolutions. At present, the null hypotheses are classical so that they assume there are no long-range statistical dependencies and that the probabilities are thin-tailed (i.e. exponential). However it is seen that cascades involve long-range dependencies and (typically) have fat-tailed (algebraic) distributions in which extreme events occur much more frequently and can persist for much longer than classical theory would allow. Indeed, the problem of statistical 'outliers' may generally be a consequence of the failure of highly variable cascade data to fit into relatively homogeneous, regular, classical geostatistical frameworks.

1.4 Applications of Nonlinear Dynamics and Chaos for Weather Prediction: Current Status

At present, the signatures of deterministic chaos, namely the fractal geometrical structure concomitant with *1/f* noise, have been conclusively identified in model and real-world atmospheric flows, and the fractal dimension of the strange attractor traced by atmospheric flows has been estimated with recently developed numerical algorithms (Grassberger and Procaccia 1983; Kerr 1989), which use the time series data of meteorological parameters, e.g. rainfall, temperature, wind speed, etc. However, such estimations of the fractal dimension have not helped resolve the problem of the formulation of a simple closed set of governing equations for atmospheric flows (Tsonis and Elsner 1989; Pool 1989a, b; Tsonis 1989; Andrews and Read 1989) mainly because the basic physics of deterministic chaos is not yet identified. A complete review of applications of concepts in *nonlinear dynamics and chaos* in atmospheric sciences has been given by Zeng et al. (1993).

During the past three decades, Lovejoy and his group (Lovejoy and Schertzer 2010) have done extensive observational and theoretical studies of fractal nature of atmospheric flows and emphasize the urgent need to formulate and incorporate quantitative theoretical concepts of fractals in mainstream classical meteorological theory. The empirical analyses summarized by Lovejoy and Schertzer (2010) directly demonstrate the strong scale dependencies of many atmospheric fields, showing that they depend in a power-law manner on the space-timescales over which they are measured. In spite of intense efforts over more than 50 years, analytic approaches have been surprisingly ineffective at deducing the statistical properties of turbulence. Atmospheric science labours under the misapprehension that its basic science issues have long been settled and that its task is limited to the application of known laws—albeit helped by ever larger quantities of data themselves processed in evermore powerful computers and exploiting ever more sophisticated algorithms. Conclusions about anthropogenic influences on the atmosphere can only be drawn with respect to the null hypothesis, i.e. one requires a theory of the natural variability, including knowledge of the probabilities of the extremes at various resolutions. At present, the null hypotheses are classical so that they assume there are no long-range statistical dependencies and that the probabilities are thin-tailed (i.e. exponential). However observations show that cascades involve long-range dependencies and (typically) have fat-tailed (algebraic) distributions in which extreme events occur much more frequently and can persist for much longer than classical theory would allow (Lovejoy and Schertzer 2010).

Dessai and Walter (2000) argue that there is enough evidence, to show that complexity and its theory of self-organized criticality (SOC) have considerable potential to increase our understanding of the atmospheric sciences and emphasized the urgent need to incorporate fundamental concepts of SOC in atmospheric as follows. Meteorologists and climatologists have largely ignored SOC. Although large power events are comparatively rare, events can and do happen on all scales, with no different mechanism needed to explain the rare large events than that which explains the smaller, more common ones (Bak 1996). In the atmospheric sciences there has been little application of what some have considered the leading candidate for a unified theory of complexity, namely, SOC. In his review of complexity and climate, Rind (1999) concludes climate, like weather will likely always be complex: '*determinism in the midst of chaos, unpredictability in the midst of understanding*'. Rind (1999) warns that it is still not known if complexity is relevant to climate science. Theories of complexity, such as SOC, have been underrepresented in the atmospheric sciences because of their 'soft science' character (Dessai and Walter 2000). Atmospheric sciences have historically developed from centuries of advancement in the hard sciences, such as physics, mathematics and chemistry, etc. It would have been unlikely to see a quick transition from the classical reductionist and reproducible science approach towards an abstract, holistic and probabilistic complex science. Proof of this is the fact that only a small number of scientists have cited the few applications of these theories in the atmospheric sciences, e.g. Vattay and Harnos (1994), Lovejoy and Schertzer (2010 and all references therein) conclude that the multifractal approach yields a convenient framework for the analysis and simulation of highly nonlinear meteorological fields over a wide range of scales and intensities and Selvam (1990, 2005, 2007, 2009, 2013) has developed a cell dynamical system model (general systems theory) for self-organized criticality (Dessai and Walter 2000).

Tsonis et al. (2006) have recently applied the concept of networks for the observed scale-free pattern for atmospheric flows as follows. Advances in understanding coupling in complex networks offer new ways of studying the collective behavior of interactive systems and already have yielded new insights in many areas of science. From this initial application of networks to climate it appears that atmospheric fields can be thought of as a network of interacting points whose collective behavior may exhibit properties of small world networks. This ensures the efficient transfer of information. In addition, the scale-free architectures guarantee stability. Furthermore, supernodes in the network identify teleconnection patterns. As was demonstrated in Tsonis (2004), these teleconnections are not static phenomena, but their spatiotemporal variability is affected by large (global) changes. Complex networks describe many natural and social dynamical systems, and their study has revealed interesting mechanisms underlying their function. The novelty of networks is that they bring out topological/geometrical aspects that are related to the physics of the dynamical system in question, thus providing a new and innovative way to treat and investigate nonlinear systems and data. While several advances have been made, this area is still young and the future is wide open. This introductory paper presented some fundamental aspects of networks and some preliminary results of the application of networks to climatic data, which indicate that networks delineate some key features of the climate system. This suggests that networks have the potential to become a new and useful tool in climate research Tsonis et al. (2006).

1.4.1 *Space-Time Cascade Model for Fractal Fluctuations in Atmospheric Flows*

Lovejoy and Schertzer (2010), the pioneers in the study of nonlinear dynamical characteristics of atmospheric flows have proposed a space-time cascade model for a realistic simulation of weather and climate as summarized in the following. In spite of the unprecedented quantity and quality of meteorological data and numerical models, there is still no consensus about the atmosphere's elementary statistical properties as functions of scale in either time or in space. The proposed model is a new synthesis based on (a) advances in the last 25 years in nonlinear dynamics, (b) a critical reanalysis of empirical aircraft and vertical sonde data, (c) the systematic scale by scale, space-time exploitation of high-resolution remotely sensed data and (d) the systematic reanalysis of the outputs of numerical models of the atmosphere including reanalyses, (e) a new turbulent model for the emergence of the climate from 'weather' and climate variability. Lovejoy and Schertzer (2010) conclude that Richardson's old idea of scale by scale simplicity—today embodied in multiplicative cascades—can accurately explain the statistical properties of the atmosphere and its models over most of the meteorologically significant range of scales, as well as at least some of the climate range (Lovejoy et al. 2009). The resulting space-time cascade model combines these nonlinear developments with modern statistical analyses, it is based on strongly anisotropic and intermittent generalizations of the classical turbulence laws of Kolmogorov, Corrsin, Obukhov, and Bolgiano.

Lovejoy and Schertzer (2010) have given an overview of a body of work carried out over the last 25 years aiming at a scale-by-scale understanding of the space-time statistical structure of the atmosphere and its models. The proposed new synthesis would not be possible without technologically driven revolutions in both data quantity and quality as well as in numerical modelling and data processing. Also key for this synthesis are advances in our understanding of nonlinear dynamics (especially cascades, multifractals, and their anisotropic extensions), and in the corresponding data analysis techniques. Although there are many gaps to fill, it is remarkable that a relatively simple picture of the atmosphere as a system of interacting anisotropic cascades seems to be consistent with some of the largest and highest quality satellite, lidar, dropsonde, and aircraft campaigns to date collectively measuring passive and active radiances over the long and short wave regimes, as well as in situ wind, temperature, humidity, potential temperature, pressure, and other variables. It also holds remarkably well for reanalyses and other numerical models of the atmosphere. It leads to a natural distinction between the weather and climate and successively predicts the transition to the climate at approximately equal to 10 days as a dimensional transition from a weather system (where both long-range space and time correlations are important) to a climate system dominated by long–range temporal correlations.

Lovejoy and Schertzer (2010) conclude that, in any case, some coherent picture is urgently needed to replace the ageing and untenable (but still dominant!) 2D isotropic/3D isotropic turbulence model.

1.4.2 General Systems Theory for Fractal Space-Time Fluctuations in Atmospheric Flows

Selvam (1990, 2005, 2007, 2009, 2011, 2013) has recently developed general systems theory for fractal space-time fluctuations based on the concept that the larger scale fluctuation can be visualized to emerge from the space-time averaging of enclosed small-scale fluctuations, thereby generating a hierarchy of self-similar fluctuations manifested as the observed eddy continuum in power spectral analyses of fractal fluctuations. Such a concept results in inverse power-law form incorporating the golden mean τ for the space-time fluctuation pattern and also for the power spectra of the fluctuations. The predicted distribution is close to the Gaussian distribution for small-scale fluctuations, but exhibits *fat long tail* for large-scale fluctuations. The general systems theory, originally developed for turbulent fluid flows, provides universal quantification of physics underlying fractal fluctuations and is applicable to all dynamical systems in nature independent of its physical, chemical, electrical, or any other intrinsic characteristic.

Macroscale coherent structures emerge by space-time integration of microscopic domain fluctuations in fluid flows. Such a concept of the autonomous growth of atmospheric eddy continuum with ordered energy flow between the scales is analogous to Prigogine's (1988) concept of the spontaneous emergence of order and organization out of apparent disorder and chaos through a process of self-organization.

The problem of emergence of macroscopic variables out of microscopic dynamics is of crucial relevance in biology (Vitiello 1992). Biological systems rely on a combination of network and the specific elements involved (Kitano 2002). The notion that membership in a network could confer stability emerged from Ludwig von Bertalanffy's description of general systems theory in the 1930s and Norbert

Wieners description of cybernetics in the 1940s. General systems theory focused in part on the notion of flow, postulating the existence and significance of flow equilibria. In contrast to Cannon's concept that mechanisms should yield homeostasis, general systems theory invited biologists to consider an alternative model of homeodynamics in which nonlinear, non-equilibrium processes could provide stability, if not constancy (Buchman 2002). The cell dynamical system model for coherent pattern formation in turbulent flows (Selvam 1990, 2005, 2007) may provide a general systems theory for biological complexity. General systems theory is a logical mathematical field, the subject matter of which is the formulation and deduction of those principles which are valid for 'systems' in general, whatever the nature of their component elements or the relations or 'forces' between them (Von Bertalanffy 1968; Peacocke 1989; Klir 1993).

In summary, the model predicts the following: (i) The eddy continuum consists of an overall logarithmic spiral trajectory with the quasiperiodic *Penrose* tiling pattern for the internal structure. (ii) The successively larger eddy space-time scales follow the Fibonacci number series. (iii) The probability distribution P of fractal domains for the nth step of eddy growth is equal to τ^{-4n} where τ is the golden mean equal to $(1 + \sqrt{5})/2$ ($\approx$1.618). The probability distribution P is close to the statistical normal distribution for n values less than 2 and greater than normal distribution for n more than 2, thereby giving a *fat, long tail.* (iv) The probability distribution P also represents the relative eddy energy flux in the fractal domain. The square of the eddy amplitude (variance) represents the eddy energy and therefore the eddy probability density P. Such a result that the additive amplitudes of eddies when squared represent probabilities, is exhibited by the subatomic dynamics of quantum systems such as the electron or proton (Maddox 1988, 1993; Rae 1988). Therefore fractal fluctuations are signatures of quantum-like chaos in dynamical systems. (v) The universal algorithm for self-organized criticality is expressed in terms of the universal *Feigenbaum*'s constants (Feigenbaum 1980) a and d as $2a^2 = \pi d$ where the fractional volume intermittency of occurrence πd contributes to the total variance $2a^2$ of fractal structures. (vi) The *Feigenbaum*'s constants are expressed as functions of the *golden mean.* The probability distribution P of fractal domains is also expressed in terms of the Feigenbaum's constants a and d.

The model predicted inverse power-law distribution has been identified in time series of meteorological parameters (Selvam et al. 1992; Selvam and Fadnavis 1998; Joshi and Selvam 1999; Selvam 2011).

1.5 Conclusions

The summit statement of the climate prediction project (Shukla et al. 2009) emphasizes the need for realistic climate/weather prediction as follows. Considerably improved predictions of the changes in the statistics of regional climate, especially of extreme events and high-impact weather, are required to assess the impacts of climate change and variations. Investing today in climate science will lead to significantly reduced costs of coping with the consequences of climate change tomorrow. Despite tremendous progress in climate modelling and the capability of high-end computers in the past 30 years, our ability to provide robust estimates of the risk to society, particularly from possible catastrophic changes in regional climate, is constrained by limitations in computer power and scientific understanding. To estimate the quality of a climate prediction requires an assessment of how accurately we know and understand the current state of natural climate variability, with which anthropogenic climate change interacts (Shukla et al. 2009).

Numerical weather/climate prediction models do not give realistic forecasts (see Sect. 3.3) because of the following inherent limitations: (1) the continuum dynamical system such as atmospheric flows is computed as a discrete dynamical system with implicit assumption of subgrid—scale homogeneity (2) model approximations and arbitrary assumptions (3) the governing equations do not incorporate the dynamical interactions and coexistence of the complete spectrum of turbulent fluctuations which form an integral part of the large coherent weather systems (Shaffee and Shaffee 1987; Lovejoy and Schertzer 1986a, b, 2010; Selvam 1988, 1994, 2007, 2013) (4) binary number representation in digital computers precludes exact number representation at the data input stage itself (5) round-off error of finite precision computer arithmetic magnifies exponentially with time the above uncertainities and gives unrealistic solution (Beck and Roepstorff 1987; Selvam 1993). Selvam (1993), in particular has shown that round-off error approximately doubles for each iteration of finite precision iterative computations and enters the mainstream computation within 50 iterations and thereafter the computed solution gives only the round-off error growth structure. Numerical weather/climate prediction models incorporate thousands of iterative computations in numerical integration schemes and therefore the model solutions will only represent the dynamical evolution of round-off error growth.

The accurate modelling of weather/climate phenomena therefore requires alternative concepts and computational techniques. Theoretical concepts and analytical techniques developed so far in the multidisciplinary new science of nonlinear dynamics and chaos have to be adapted and incorporated in classical meteorological theory for realistic prediction of weather phenomena.

References

Abarbanel, H.D.I.: Analysis of Observed Chaotic Data. Springer, New York (1996)

Andrews, D., Read, P.: Chaos and predictability in meteorology. Phys. World **2**, 20 (1989)

Andriani, P., McKelvey, B.: Beyond Gaussian averages: redirecting management research toward extreme events and power laws. J. Int. Bus. Stud. **38**, 1212–1230 (2007)

Arber, A.: The Natural Philosophy of Plant Form. Cambridge University Press, London (1950)

Arecchi, F.T.: Optical morphogenesis: pattern formation and competition in nonlinear optics. Phys. D **86**, 297–322 (1995)

Baake, M.: Quasicrystals: An introduction to structure, physical properties and applications. In: Suck, J.B. et al. (eds.) Springer, Berlin (2002)

Back, C.H., Wurch, Ch., Avateriaus, A., Ramsperger, U., Maier, U., Pescia, D.: Experimental confirmation of universality for a phase transition in two dimensions. Nature **378**, 597–600 (1995)

Bak, P.: How nature works: the science of self-organized criticality. Copernicus, New York (1996). 212 pp

Bak, P., Chen, K.: The physics of fractals. Phys. D **38**, 5–12 (1989)

Bak, P., Chen, K.: Self-organized criticality. Sci. Am. **Jan.**, 26–33 (1991)

Bak, P., Tang, C., Wiesenfeld K.: Self-organized criticality: an explanation of 1/f noise. Phys. Rev. Lett. **59**, 381–384 (1987)

Bak, P.C., Tang, C., Wiesenfeld, K.: Self-organized criticality. Phys. Rev. A. **38**, 364–374 (1988)

Bassingthwaighte, J.B., Beyer, R.P.: Fractal correlations in heterogeneous systems. Phys. D **53**, 71–84 (1991)

Beck, C., Roepstorff, G.: Effects of phase space discretization on the long time behaviour of dynamical systems. Physica **25D**, 173–180 (1987)

Besicovitch, A.E.: On linear sets of points of fractional dimension. Math. Ann. **101**, 161–193 (1929)

Blank, M.: Pathologies generated by round-off in dynamical systems. Phys. D **78**, 93–114 (1994)

Buchanan, M.: One law to rule them all. New Sci. **8 Nov.**, 30–35 (1997)

Buchman, T.G.: The community of the self. Nature **420**, 246–251 (2002)

Buchanan, M.: Power laws and the new science of complexity management. Strat. Bus. **34**, 70–79 (2004)

Canavero, F.G., Einaudi, F.: Time and space variability of atmospheric processes. J. Atmos. Sci. **44**(12), 1589–1604 (1987)

Clauset, A., Shalizi, C.R., Newman, M.E.J.: Power-law distributions in empirical data (2007). arXiv:0706.1062v1 [physics.data-an] 2007

Corless, R.M., Frank, G.W., Monroe, J.G.: Chaos and continued fractions. Phys. D **46**, 241–253 (1990)

Csete, M.E., Doyle, J.C.: Reverse engineering of biological complexity. Science **295**, 1664–1669 (2002)

Dangelmayr, G.: In: Guettinger, W., Dangelmayr, G. (eds.) The Physics of Structure Formation. Springer, Berlin (1987)

Deering, W., West, B.J.: Fractal physiology. IEEE Eng. Med. Biol. **June**, 40–46 (1992)

Dennin, M., Ahlers, G., Cannell, D.S.: Spatiotemporal chaos in electro-convection. Science **272**, 388–396 (1996)

Dessai, S., Walter, M.E.: Self-organized criticality and the atmospheric sciences: selected review, new findings and future directions, pp. 14. *NCAR Extreme Events workshop (2000)*, Department of Environmental Studies, University of Colorado, Boulder, CO, USA http://www.esig.ucar.edu/extremes/papers/walter.PDF (2000)

Dewan, E.M., Good, R.E.: Saturation and the universal spectrum for vertical profiles of horizontal scalar winds in the atmosphere. J. Geophys. Res. (Atmos.) **91**, 2742–2748 (1986)

Dewan, E.M., Grossbard, N., Good, R.E., Brown, J.: Power spectral densities of zonal and meridional winds in the stratosphere. Phys. Scr. **37**, 154–157 (1988)

Dewdney, A.K.: Computer recreations. Sci. Am. **255**, 14–23 (1986)

DiVincenzo, D.P.: Perfect quasicrystals? Nature **340**, 504 (1989)

Eymard, L.: Convective organization in a tropical boundary layer: an interpretation of doppler radar observations using Asai's model. J. Atmos. Sci. **42**, 2844–2855 (1985)

Feigenbaum, M.J.: Quantitative universality for a class of nonlinear transformations. J. Stat. Phys. **19**(1), 25–52 (1978)

Feigenbaum, M.J.: Universal behavior in nonlinear systems. Los Alamos Sci. **1**, 4–27 (1980)

Freeman, G.R.: Introduction. In: Freeman, G.R. (ed.) Kinetics of Nonhomogenous Processes, pp. 1–18. Wiley, New York (1987)

Freeman, G.R.: KNP89: Kinetics of nonhomogenous processes (KNP) and nonlinear dynamics. Can. J. Phys. **68**, 655–659 (1990)

Frisch, U., Orszag, S.A.: Turbulence: challenges for theory and experiment. Phys. Today **Jan.**, 24–32 (1990)

Fritts, D.C., Chou, H.G.: An investigation of the vertical wavenumber and frequency spectra of gravity wave motions in the lower stratosphere. J. Atmos. Sci. **44**, 3610–3624 (1987)

Ghashghaie, S., Breymann, Peinke, J., Talkner, P., Dodge, Y.: Turbulent cascades in foreign exchange markets. Nature **381**, 767–770 (1996)

Gleick, J.: Chaos: Making a New Science. Viking, New York (1987)

Goldberger, A.L., Rigney, D.R., West, B.J.: Chaos and fractals in human physiology. Sci. Am. **262**(2), 42–49 (1990)

Grassberger, P., Procaccia, I.: Characterization of strange attractors. Phy. Rev. Lett. **50**, 346–349 (1983)

Grebogi, C., Ott, E., Yorke, J.A.: Round-off-induced periodicity and the correlation dimension of chaotic attracts. Phys. Rev. A: Gen. Phys. **38**, 3688–3692 (1988)

Greene, W.H.: Econometric Analysis, 5th edn. Prentice-Hall, Englewood Cliffs, NJ (2002)

Guckenheimer, J., Holmes, P.J.: Nonlinear Oscillations, Dynamical Systems and Bifurcation of Vector Fields. Springer, New York (1983)

Gutenberg, B., Richter, R.F.: Frequency of earthquakes in California. Bull. Seis. Soc. Am. **34**, 185–188 (1944)

Hacker, J., Hansen, J., Berner, J., Chen, Y., Eshel, G., Hakim, G., Lazarus, S., Majumdar, S., Morss, R., Poje, A., Sheremet, V., Tang, Y., Webb, C.: Predictability. Bull. Amer. Meteorol. Soc. **86**(12), 1733–1737 (2005)

Haken, H.: Synergetics—are cooperative phenomena governed by universal priniples? Naturwissenschaften **67**, 121–128 (1980)

Haken, H.: Synergetics: an overview. Rep. Prog. Phys. **52**(7), 517–552 (1989)

Hargittai, I. (ed.): Fivefold Symmetry. World Scientific, Singapore (1992)

Hargittai, I., Pickover, C.A. (eds.): Spiral Symmetry. World Scientific, Singapore (1992)

Hausdorff, F.: Dimension und ausseres mass. Math. Ann. **29**, 157–179 (1919)

Hooge, C., Lovejoy, S., Schertzer, D., Pecknold, S., Malouin, J.F., Schmitt, F.: Multifractal phase transitions: the origin of self-organized criticality in earthquakes. Nonlinear Process. Geophys. **1**, 191–197 (1994)

Hurst, H.E.: Long-term storage capacity of reservoirs. Trans. Amer. Soc. Civ. Engrs. **116**, 770–808 (1951)

Hurst, H.E., Black, R.P., Simaiki, Y.M.: Long-term storage: an experimental study. Constable, London (1965)

Insinnia, E.M.: Synchronicity and coherent excitations in microtubules. Nanobiology **1**(2), 191–208 (1992)

Jean, R.V.: Phyllotaxis: a systemic study in plant morphogenesis. Cambridge University Press, New York (1994)

Joshi, R.R., Selvam, A.M.: Identification of self-organized criticality in atmospheric low frequency variability. Fractals Complex Geom. Patterns Scaling Nat. Soc. **4**, 421–425 (1999)

Jurgen, H., Peitgen, H.-O., Saupe, D.: The language of fractals. Sci. Am. **263**, 40–49 (1990)

Kadanoff, L.P.: Turbulent excursions. Nature **382**, 116–117 (1996)

Kalnay, E.: Atmospheric Modeling, Data Assimilation and Predictability, 341 pp. Cambridge University Press (2003)

Kappraff, J.: The relationship between mathematics and mysticism of the golden mean through history. In: Hargittai, I. (ed.) Fivefold Symmetry, pp. 33–65. World Scientific, Singapore (1992)

Kerr, R.A.: Telling weathermen when to worry. Science **244**, 1137–1139 (1989)

Kerr, R.A.: Climate modelling's fudge factor comes under fire. Science **265**, 1528–1529 (1994)

Kitano, H.: Computational systems biology. Nature **420**, 206–210 (2002)

Kitano, H.: Biological robustness. Nat. Rev. Genet. **5**, 826–837 (2004)

Klir, G.J.: Systems science: a guided tour. J. Biol. Syst. **1**(1), 27–58 (1993)

Kolmogorov, A.N.: The local structure of turbulence in incompressible liquids for very high Reynolds numbers. C. R. Russ. Acad. Sci. **30**, 301–305 (1941)

Kolmogorov, A.N.: A refinement of previous hypotheses concerning the local structure of turbulence in a viscous inhomogeneous fluid at high Reynolds number. J. Fluid Mech. **13**, 82–85 (1962)

Kushnir, Y., Wallace, J.M.: Low-frequency variability in the northern hemisphere winter: geographical distribution, structure and time-scale dependence. J. Atmos. Sci. **46**, 3122–3142 (1989)

Lau, K.M., Peng, L., Sui, C.H., Nakazwa, T.: Dynamics of super cloud clusters, westerly wind bursts, 30-60 day oscillations and ENSO- An unified view. J. Meteorol. Soc. Jpn **67**, 205–219 (1989)

Levich, E.: Certain problems in the theory of developed hydrodynamic turbulence. Phys. Rep. **151** (3, 4), 129–238 (1987)

Levine, D., Steinhardt, J.: Quasicrystals: a new class of ordered structures. Phys. Rev. Letts. **53** (26), 2477–2480 (1984)

Levy, P.: Theorie de l'addition des variables aleatoires. Gauthier-Villiers, Paris (1937)

Li, W.: A bibliography on 1/*f* noise. http://www.nslij-genetics.org/wli/1fnoise (2007)

Liebovitch, L.S., Scheurle, D.: Two lessons from fractals and chaos. Complexity **5**(4), 34–43 (2000)

Lighthill, J.: The recently recognized failure of predictability in Newtonian dynamics. Proc. R. Soc. Lond. A **407**, 35–50 (1986)

Lilly, D.K.: On the structure, energies and propagation of rotating convective storms: Part II, Helicity and storm stabilization. J. Atmos. Sci. **43**, 126–140 (1986)

Lilly, D.K.: Two-dimensional turbulence generated by energy sources at two scales. J. Atmos. Sci. **46**, 2026–2030 (1989)

Liu, S.H.: Formation and anomalous properties of fractals. IEEE Eng. Med. Biol. **June**, 28–39 (1992)

Lord, E.A.: Quasicrystals and penrose patterns. Curr. Sci. **61**(5), 313–319 (1991)

Lorenz, E.N.: Deterministic nonperiodic flow. J. Atmos. Sci. **30**, 130–141 (1963)

Lorenz, E.N.: A study of the predictability of a 28-variable atmospheric model. Tellus **17**, 321–333 (1965)

Lorenz, E.N.: The predictability of a flow which possesses many scales of motion. Tellus **21**, 289–307 (1969)

Lorenz, E.N.: Irregularity: a fundamental property of the atmosphere. Tellus **36A**, 98–110 (1984)

Lorenz, E.N.: Atmospheric models as dynamical systems. In: Shlesinger, M.F., Crawley, R., Saenz, A.W., Zachary, W. (eds.) Perspectives in nonlinear dynamics, pp. 1–17. World Scientific, Singapore (1986)

Lorenz, E.N.: Computational chaos—a prelude to computational instability. Phys. D **35**, 299–317 (1989)

Lorenz, E.N.: Can chaos and intransitivity lead to interannual variability? Tellus **42A**, 378–389 (1990)

Lorenz, E.N.: Dimension of weather and climate attractors. Nature **353**, 241–244 (1991)

Lovejoy, S., Schertzer, D.: Towards a new synthesis for atmospheric dynamics: space-time cascades. Atmos. Res. (in press) (2010). doi:10.1016/j.atmosres.2010.01.004. http://physics.mcgill.ca/~gang/-/cascades.summary.final.16.11.9.pdf

Lovejoy, S., Schertzer, D.: Scale invariance, symmetries, fractal and stochastic simulations of atmospheric phenomena. Bull. Amer. Meteorol. Soc. **67**, 21–32 (1986a)

Lovejoy, S., Schertzer, D.: Scale invariance in climatological temperatures and the local spectral plateau. Ann. Geophys. **4B**, 401–410 (1986b)

Lovejoy, S., Schertzer, D., Allaire, V., Bourgeois, T., King, S., Pinel, J., Stolle, J.: Atmospheric complexity or scale by scale simplicity? Geophys. Res. Letts. **36**, L01801, 6 pp (2009)

Maddox, J.: Licence to slang Copenhagen? Nature **332**, 581 (1988)

Maddox, J.: Long-range correlations within DNA. Nature **358**, 103 (1992)

Maddox, J.: Can quantum theory be understood? Nature **361**, 493 (1993)

Mandelbrot, B.B.: On the geometry of homogenous turbulence with stress on the fractal dimension of the iso-surfaces of scalars. J. Fluid Mech. **72**, 401–416 (1975)
Mandelbrot, B.B.: Fractals: form, chance and dimension. Freeman, San Francisco (1977)
Mandelbrot, B.B.: The fractal geometry of nature. Freeman, W.H (1983). 468 pp
Mandelbrot, B.B., Wallis, J.R.: Some long-run properties of geophysical records. Water Resour. Res. **5**, 321–340 (1969)
Mantegna, R.N., Stanley, H.E.: Scaling behaviour in the dynamics of an economic index. Nature **376**, 46–49 (1995)
Mason, B.J.: Numerical weather prediction. Proc. R. Soc. Lond. A **407**, 51–60 (1986)
Mayer-Kress: Fractals. Phys. Bull. **39**, 357–359 (1988)
Mccauley, J.L.: An introduction to nonlinear dynamics and chaos theory. Phy. Scr. T **20**, 1–57 (1988)
Milotti, E.: *1/f* noise: a pedagogical review (2002). http://arxiv.org/abs/physics/0204033
Mintmire, J.W.: Fullerene formation and annealing. Science **272**, 45–46 (1996)
Monin, A.S., Yaglom, A.M.: Statistical hydrodynamics, vols. 1 and 2. MIT Press, Cambridge (1975)
Muller, A., Beugholt, C.: The medium is the message. Nature **383**, 296–297 (1996)
Nelson, D.R.: Quasicrystals. Sci. Am. **255**, 42–51 (1986)
Newman, M.: The power of design. Nature **405**, 412–413 (2000)
Nicolis, G., Prigogine, I.: Self-Organization in Non Equilibrium Systems. Wiley, New York (1977)
Omori, F.: On the aftershocks of earthquakes. J. Coll. Sci. Imp. Univ. Tokyo **7**, 111–200 (1895)
Palmer, T.N., Doblas-Reyes, F.J., Hagedorn, R., Weisheimer, A.: Probabilistic prediction of climate using multi-model ensembles: from basics to applications. Phil. Trans. R. Soc. B **360**, 1991–1998 (2005)
Peacocke, A.R.: The Physical Chemistry of Biological Organization. Clarendon Press, Oxford, U. K. (1989)
Peitgen, H.-O., Richter, P.H., Saupe, D.: Chaos and Fractals: New Frontiers in Science, p. 971. Springer, New York (1992)
Peng, C.K., Buldyrev, S.V., Goldberger, A.L., Havlin, S., Sciortino, F., Simons, M., Stanley, H.E.: Long-range correlations in nucleotide sequences. Nature **356**, 168–170 (1992)
Penrose, R.: Pentaplexity. Math. Intell. **2**(1), 32–37 (1979)
Penrose, R.: The role of aesthetics in pure and applied mathematical research. Bull. Inst. Math. Appl. **10**, 266–271 (1974), reprinted In: Steinhardt, P.J., Ostlund, S. (eds.) The Physics of Quasicrystals. World Scientific, Singapore (1987)
Poincare, H.: Acta Math. **13**, 5 (1890)
Poincare, H.: Les Methodes Nouvelle de la Mecannique Celeste. Gautheir-Villars, Paris (1892)
Pool, R.: Is something strange about the weather? Science **243**, 1290 (1989a)
Pool, R.: Chaos theory: how big an advance? Science **245**, 26 (1989b)
Prigogine, I.: From Being to Becoming. Freeman, San Francisco, CA, USA (1980)
Prigogine, I., Stengers, I.: Order Out of Chaos, 3rd edn. Fontana Paperbacks, London (1988)
Rae, A.: Quantum-Physics: Illusion or Reality?. Cambridge University Press, New York (1988)
Reinhold, B.: Weather regimes: the challenge of extended range forecasting. Science **235**, 437–441 (1987)
Rhodes, C.J., Anderson, R.M.: Power laws governing epidemics in isolated populations. Nature **381**, 600–602 (1996)
Richardson, L.F.: Weather Prediction by Numerical Process. Dover, Mineola, N. Y. (1965)
Richardson, L.F.; The problem of contiguity: an appendix to statistics of deadly quarrels. In: Von Bertalanffy, L., Rapoport, A., (eds.) General Systems—Year book of the Society for General Systems Research, pp. 139–187, Ann Arbor, MI (1960)
Rind, D.: Complexity and climate. Science **284**, 105–107 (1999)
Roebber, P.J., Tsonis, A.A.: A method to improve prediction of atmospheric flow transitions. J. Atmos. Sci. **62**, 3818–3824 (2005)

Schepers, H.E., Van Beek, J.H.G.M., Bassingthwaighte, J.B.: Four methods to estimate the fractal dimension from self-affine signals. IEEE Eng. Med. Biol. **June**, 57–71 (1992)
Schertzer, D., Lovejoy, S.: Multifractal generation of self-organized criticality. In: Novak, M.M. (ed.) Fractals in the Natural and Applied Sciences (A-41), pp. 325–339. Elsevier Science B. V. (North-Holland) (1994)
Schertzer, D., Lovejoy, S.: Generalized scale invariance and multiplicative processes in the atmosphere. Pure. Appl. Geophys. **130**, 57–81 (1989)
Schertzer, D., Lovejoy, S.: Nonlinear geodynamical variability, multiple singularities, universality and observables. In: Schertzer, D., Lovejoy, S. (eds.) Scaling, Fractals and Nonlinear Variability in Geophysics, pp. 41–82. Kluwer Academic, Norwell, Mass (1991)
Schroeder, M.: Fractals, Chaos and Power-laws. W. H. Freeman and Co., N. Y (1991)
Selvam, A.M.: The dynamics of deterministic chaos in numerical weather prediction models. In: Proceedings of the American Meteorological Society 8th Conference. Numerical Weather Prediction, February 1988, Baltimore, MD (1988)
Selvam, A.M.: Deterministic chaos, fractals and quantumlike mechanics in atmospheric flows. Can. J. Phys. **68**, 831–841 (1990). http://xxx.lanl.gov/html/physics/0010046
Selvam, A.M.: Universal quantification for deterministic chaos in dynamical systems. Appl. Math. Model. **17**, 642–649 (1993). http://xxx.lanl.gov/html/physics/0008010
Selvam, A.M.: The physics of deterministic chaos: implications for global climate model predictions. In: Proceedings of the International Conference on Global Climate Change: Science, Policy and Mitigation Strategies, April 5–8, 1994, pp. 412–417, Phoenix, Arizona (1994)
Selvam, A.M.: A general systems theory for chaos, quantum mechanics and gravity for dynamical systems of all space-time scales. Electromagn. Phenom. **5**(2, 15), 160–176 (2005). http://arxiv.org/pdf/physics/0503028; http://www.emph.com.ua/15/selvam.htm
Selvam, A.M.: Chaotic Climate Dynamics. Luniver Press, UK (2007)
Selvam, A.M.: Fractal fluctuations and statistical normal distribution. Fractals **17**(3), 333–349 (2009). http://arxiv.org/pdf/0805.3426
Selvam, A.M.: Signatures of universal characteristics of fractal fluctuations in global mean monthly temperature anomalies. J. Syst. Sci. Complex. **24**(1), 14–38 (2011). http://arxiv.org/pdf/0808.2388v1
Selvam, A.M.: Nonlinear dynamics and chaos: applications in atmospheric sciences. J. Adv. Math. Appl. **1**, 1–24 (2013)
Selvam, A.M., Fadnavis, S.: Signatures of a universal spectrum for atmospheric inter-annual variability in some disparate climatic regimes. Meteorol. Atmos. Phys.**66**, 87–112 (1998). http://xxx.lanl.gov/abs/chao-dyn/9805028
Selvam, A.M., Pethkar, J.S., Kulkarni, M.K.: Signatures of a universal spectrum for interannual variability in rainfall time series over the Indian region. Int. J. Climatol. **12**, 137–152 (1992)
Shaffee, A., Shaffee, S.: Implication of the spatial finiteness of mesoscale meteorological fields. Phys. Rev. A **35**, 892–896 (1987)
Shepherd, T.G.: A spectral view of nonlinear fluxes and stationary-transient interaction in the atmosphere. J. Atmos. Sci. **44**, 1166–1179 (1987)
Shlesinger, M.F., West, B.J., Klafter, J.: Levy dynamics of enhanced diffusion: application to turbulence. Phy. Rev. Lett. **58**(110), 1100–1103 (1987)
Shukla, J., Hagedorn, R., Hoskins, B., Kinter, J., Marotzke, J., Miller, M., Palmer, T.N., Sungo, J.: Strategies: revolution in climate prediction is both necessary and possible: a declaration at the world modelling summit for climate prediction. Bull. Am. Meteor. Soc. **90**(2), 175–178 (2009)
Skinner, J.E.: Low dimensional chaos in biological systems. Bio/Technology **12**, 596–600 (1994)
Sreenivasan, K.R.: Fractals and multifractals in turbulence. Annu. Rev. Fluid Mech. **23**, 539–600 (1991)
Stanley, H.E.: Powerlaws and universality. Nature **378**, 554–555 (1995)
Stanley, M.H.R., Amaral, L.A.N., Buldyrev, S.V., Havlin, S., Leschhorn, H., Maass, P., Salinger, M.A., Stanley, H.E.: Can statistical physics contribute to the science of economics? Fractals **4** (3), 415–425 (1996a)

Stanley, H.E., Amaral, L.A.N., Buldyrev, S.V., Goldberger, A.L., Havlin, S., Hyman, B.T., Leschhorn, H., Maass, P., Makse, H.A., Peng, C.-K., Salinger, M.A., Stanley, M.H.R., Vishwanathan, G.M.: Scaling and universality in living systems. Fractals **4**(3), 427–451 (1996b)

Steinhardt, P.: Crazy crystals. New Sci. **25 Jan**, 32–35 (1997)

Stevens, P.S.: Patterns in Nature. Little, Brown and Co., Inc., Boston, USA (1974)

Stewart, I.: Where do nature's patterns come from ? Nature **135**, 14 (1992a)

Stewart, I.: Warning-handle with care ! Nature **355**, 16 (1992b)

Stewart, I.: Daisy, daisy, give your answer do. Sci. Am. **272**, 76–79 (1995)

Stoddart, F.: Unnatural product synthesis. Nature **334**, 10 (1988)

Strathmann, R.R.: Testing size abundance rules in a human exclusion experiment. Science **250**, 1091 (1990)

Suki, B., Barabasi, A., Hantos, Z., Petak, F., Stanley, H.E.: Avalanches and power-law behaviour in lung inflation. Nature **368**, 615–618 (1994)

Sun, H.H., Charef, A.: Fractal system—a time domain approach. Ann. Biomed. Eng. **18**, 597–621 (1990)

Tang, C., Bak, P.: Critical exponents and scaling relations for self organized critical phenomena. Phys. Rev. Lett. **60**, 2347–2350 (1988)

Tarasov, L.: This Amazingly Symmetrical World. Mir Publishers, Moscow (1986)

Tennekes, H.: In: Haughen, D.A. (eds.) Workshop on micrometeorology, pp. 175–216. Amer. Meteorol. Soc. Boston, USA (1973)

Tessier, Y., Lovejoy, S., Schertzer, D.: Universal multifractals: theory and observations for rain and clouds. J. Appl. Meteorol. **32**, 223–250 (1993)

Tessier, Y., Lovejoy, S., Hubert, P., Schertzer, D., Pecknold, S.: Multifractal analysis and modelling of rainfall and river flows and scaling, casual transfer function. J. Geophys. Res. **101** (D21), 26427–26440 (1996)

Thompson, P.: Uncertainty in the initial state as a factor in the predictability of large scale atmospheric flow patterns. Tellus **9**, 275–295 (1957)

Thompson, D.W.: On growth and form, 2nd edn. Cambridge University Press, Cambridge (1963)

Trenberth, K.E., Branslator, G.W., Arkin, P.A.: Origins of the 1988 North American drought. Science (Washington, D. C.) **242**, 1640–1645 (1988)

Tribbia, J., Baumhefner, D.: The reliability of improvements in deterministic short-range forecasts in the presence of initial state and modeling defi ciencies. Mon. Wea. Rev. **116**, 2276–2288 (1988)

Tsonis, A.A.: Does global warming inject randomness into the climate system? Eos Trans. Amer. Geophys. Union **85**, 361–364 (2004)

Tsonis, A.A., Elsner, J.B.: Chaos, strange attractors, and weather. Bull. Amer. Meteorol. Soc. **70** (1), 14–23 (1989)

Tsonis, A.A., Swanson, K.L., Roebber, P.J.: What do networks have to do with climate? Bull. Amer. Meteorol. Soc. **87**(5), 585–595 (2006)

Tsuda, T., Inoue, T., Fritts, D.C., VanZandt, T.E., Kato, S., Sato, T., Fukao, S.: MST Radar observations of a saturated gravity wave spectrum. J. Atmos. Sci. **46**, 2440–2447 (1989)

Turing, A.M.: The chemical basis of morphogenesis. Phil. Trans. Roy Soc. (London) **B237**, 37–52 (1952)

Ueda, Y.: Randomly transitional phenomena in the system governed by Duffing's equation. J. Stat. Phys. **20**(2), 181–196 (1979)

Van Der Zeil, A.: On the noise spectra of semiconductor noise and flicker effects. Physica **16**, 359–372 (1950)

Van Zandt, T.E.: A universal spectrum of buoyancy waves in the atmosphere. Geophys. Res. Lett. **9**, 575–578 (1982)

Vattay, G., Harnos, A.: Scaling behavior in daily air humidity fluctuations. Phys. Rev. Lett. **73**, 768–771 (1994)

Vitiello, G.: Coherence and electromagnetic fields in living matter. Nanobiology **1**, 221–228 (1992)

Von Baeyer, H.: Impossible crystals. Discover **Feb.**, 69–78 (1990)
Von Bertalanffy, L.: General Systems Theory: Foundations, Development, Applications. George Braziller, New York (1968)
Weil, J.C.: Updating applied diffusion models. J. Climat. Appl. Meteorol. **24**, 1111–1130 (1985)
Weinberg, S.: Dreams of a final theory. Vintage, pp 250 (1993)
West, B.J.: Fractal forms in physiology. Int. J. Modern Phys. B **4**(10), 1629–1669 (1990a)
West, B.J.: Physiology in fractal dimensions. Ann. Biomed. Eng. **18**, 135–149 (1990b)
West, B.J.: Comments on the renormalization group, scaling and measures of complexity. Chaos Solitons Fractals **20**, 33–44 (2004)
West, B.J., Shlesinger, M.F.: On the ubiquity of *1/f* noise. Int J. Mod. Phys. B **3**(6), 795–819 (1989)
Wilson, K.G.: Problems in physics with many scales of length. Sci. Am. **241**(2), 140–157 (1979)
Zeng, X., Pielke, R.A., Eykholt, R.: Chaos theory and its applications to the atmosphere. Bull. Amer. Meteorol. Soc. **74**(4), 631–644 (1993)
Zupanski, M., Navon, I.M.: Predictability, observations, and uncertainties in the geosciences. Bull. Amer. Meteorol. Soc. **88**(9), 1431–1433 (2006)

Chapter 2
Noise or Random Fluctuations in Physical Systems: A Review

Abstract 'Noise' or random fluctuations characterize all physical systems in nature ranging from biology, botany, physiology, meteorology, astronomy, etc. The apparently irregular or chaotic fluctuations were considered as 'noise' in all fields except in astronomy, where the fluctuations from astronomical sources were referred to as signal. Noise and fluctuation has been a field of study since 1826 with the study of Brownian motion which indirectly confirmed the existence of atoms and molecules. The measured characteristics of noise contain recognizable patterns or signal and convey useful information about the system. Statistical data analysis techniques are used to extract the signal, i.e. recognizable patterns in the apparently random fluctuations of physical systems. The analysis of data sets and broad quantification in terms of probabilities belongs to the field of statistics. Early attempts resulted in identification of the following two quantitative (mathematical) distributions which approximately fit data sets from a wide range of scientific and other disciplines of study. The first is the well-known statistical normal distribution and the second is the power-law distribution associated with the recently identified 'fractals' or self-similar characteristic of data sets in general. Abraham de Moivre, an eighteenth-century statistician and consultant to gamblers made the first recorded discovery of the normal curve of error (or the bell curve because of its shape) in 1733. The importance of the normal curve stems primarily from the fact that the distributions of many natural phenomena are at least approximately normally distributed. This normal distribution concept underlies how we analyse experimental data over the last 200 years. Most quantitative research involves the use of statistical methods presuming *independence* among data points and Gaussian 'normal' distributions. The Gaussian distribution is reliably characterized by its stable mean and finite variance. Normal distributions place a trivial amount of probability far from the mean and hence the mean is representative of most observations. Even the largest deviations, which are exceptionally rare, are still only about a factor of two from the mean in either direction and are well characterized by quoting a simple standard deviation. However, apparently rare real-life catastrophic events such as major earth quakes, stock market crashes, heavy rainfall events, etc., occur more frequently than indicated by the normal curve, i.e. they exhibit a probability distribution with a *fat tail*. Fat tails indicate a power-law pattern and interdependence.

A.M. Selvam, *Self-organiszed Criticality and Predictability in Atmospheric Flows*,
Springer Atmospheric Sciences, DOI 10.1007/978-3-319-54546-2_2

The 'tails' of a power-law curve—the regions to either side that correspond to large fluctuations—fall off very slowly in comparison with those of the bell curve. The normal distribution is therefore an inadequate model for extreme departures from the mean. For well over a century evidence had been mounting that real-world behaviour in particular, behaviour of *systems*, whether natural, social, economic, or financial does not follow normal distribution characteristics. There is increased evidence for non-normality in real-world settings and in its place an alternative distribution, namely the power-law distribution is shown to be exhibited by real-world systems in all fields of science and other areas of human interest. In this chapter, the following are discussed. (i) A brief history of the two chief quantitative methods of statistical data analysis, namely the statistical normal distribution and the power-law distribution. (ii) The association of power-law distributions with complex systems, scale invariance, self-similarity, fractals, 1/*f* noise, long-term memory, phase transitions, critical phenomena, and self-organized criticality. (iii) Current status of power-law distributions. (iv) Power-law relations (bivariate) and power-law (probability) distributions. (v) Allometric scaling and fractals. (vi) Fractals and the golden section in plant growth. (vii) Turbulent fluid flow structure, fractals, and the golden ratio (≈1.618). (viii) Fractal space-time and the golden ratio. (ix) Power-law (probability) distributions in the meteorological parameters precipitation, temperature, quaternary ice volume fluctuations and atmospheric pollution. (x) General systems theory model for self-organized criticality (SOC) in atmospheric flows with universal quantification for power-law distribution in terms of the golden ratio.

Keywords Noise and fluctuations · Power law · 1/*f* noise · Self-organized criticality (SOC) · General systems theory

2.1 Introduction

The random fluctuations in space and time are ubiquitous to physical systems in nature and are commonly known as 'noise'. Such noise or fluctuations from astronomical sources however have always been referred to as signal. The apparent 'noise' exhibits some form of regularity seen in (i) size (or duration) distribution in space (or time) and (ii) power (variance) spectrum which exhibits inverse power-law form namely 'flicker noise or 1/*f* noise' (Press 1978).

The roots of noise research trace back to the Scottish botanist Robert Brown who carried out his famous experiments in 1827, observing fluctuating pollen on the surface of a film of water. The first unsolved noise problem was to find out the origin of Brownian motion. It took over 80 years to fully solve it. Around 1905, through the work of both Smoluchowski (1906) and Einstein (1905), the problem was finally settled and presented it as a way to indirectly confirm the existence of atoms and molecules. The random fluctuations of Brownian motion are a visible manifestation of the ceaseless molecular bombardment of suspended particulates in a liquid or gas (Abbott 2001).

Noise and fluctuation has been a field of study since 1826 and now covers all fields of science and other areas of human interest. Noise is generally associated with degradation in performance, particularly in linear systems. However, it is now recognized that noise can play a constructive role in nonlinear systems, whose performance can be optimized at nonzero noise levels. Also, the measured characteristics of noise contain recognizable patterns or signal and convey useful information about the system. In recent years, noise is an intensive field of study in physical, biological, and other systems since it is recognized that depending on circumstances, noise plays a constructive or destructive role and can be utilized for enhancing required system performance (Abbott 2001).

2.2 Statistical Methods for Data Analysis

Statistical data analysis techniques are used to extract the signal, i.e. recognizable patterns in the apparently random fluctuations of physical systems.

Selvam (2009) summarized the current status of statistical methods for data analysis as follows. Dynamical systems such as atmospheric flows, stock markets, heartbeat patterns, population growth, traffic flows, etc., exhibit irregular space-time fluctuation patterns. Quantification of the space-time fluctuation pattern will help predictability studies, in particular, for events which affect day-to-day human life such as extreme weather events, stock market crashes, traffic jams, etc. The analysis of data sets and broad quantification in terms of probabilities belongs to the field of statistics. Early attempts resulted in identification of the following two quantitative (mathematical) distributions which approximately fit data sets from a wide range of scientific and other disciplines of study. The first is the well-known statistical normal distribution and the second is the power-law distribution associated with the recently identified 'fractals' or self-similar characteristic of data sets in general. In the following, a summary is given of the history and merits of the two distributions.

2.3 Statistical Normal Distribution

Historically, our present-day methods of handling experimental data have their roots about four hundred years ago. At that time scientists began to calculate the odds in gambling games. From those studies emerged the theory of probability and subsequently the theory of statistics. These new statistical ideas suggested a different and more powerful experimental approach. The basic idea was that in some experiments, random errors would make the value measured a bit higher and in other experiments random errors would make the value measured a bit lower. Combining these values by computing the average of the different experimental results would make the errors cancel and the average would be closer to the 'right' value than the result of any one experiment (Liebovitch and Scheurle 2000).

Abraham de Moivre, an eighteenth-century statistician and consultant to gamblers made the first recorded discovery of the normal curve of error (or the bell curve because of its shape) in 1733. The normal distribution is the limiting case of the binomial distribution resulting from random operations such as flipping coins or rolling dice. Serious interest in the distribution of errors on the part of mathematicians such as Laplace and Gauss awaited the early nineteenth century when astronomers found the bell curve to be a useful tool to take into consideration the errors they made in their observations of the orbits of the planets (Goertzel and Fashing 1981).

The bell curve was noticed by Gauss in the distribution of estimates of geographical measurements in the Bavarian hills. He used the curvature of Earth to improve accuracy of measurement. The distribution of estimates was found to cluster around the mean with symmetry on either side. Gauss is credited with developing the 'least squares' method for minimizing random errors in statistical inference. Along with Gauss' studies, Laplace in 1810 showed mathematically that the normal distribution follows from the central limit theorem, namely, the sum of a large number of mutually independent, identically distributed random variables is approximately normally distributed (Haldane 2012).

In probability theory, the standard Gaussian distribution arises as the limiting distribution of a large class of distributions of random variables (with suitable centering and normalization) characterized by a finite variance, which is nothing but the statement of the central limit theorem (Sornette 2012).

The importance of the normal curve stems primarily from the fact that the distributions of many natural phenomena are at least approximately normally distributed. This normal distribution concept has molded how we analyse experimental data over the last 200 years. We have come to think of data as having values most of which are near an average value, with a few values that are smaller, and a few that are larger. The probability density function, PDF(x), is the probability that any measurement has a value between x and $x + dx$. We suppose that the PDF of the data has a normal distribution. Most quantitative research involves the use of statistical methods presuming *independence* among data points and Gaussian 'normal' distributions (Andriani and McKelvey 2007). The Gaussian distribution is reliably characterized by its stable mean and finite variance (Greene 2002). Normal distributions place a trivial amount of probability far from the mean and hence the mean is representative of most observations. Even the largest deviations, which are exceptionally rare, are still only about a factor of two from the mean in either direction and are well characterized by quoting a simple standard deviation (Clauset et al. 2009). However, apparently rare real-life catastrophic events such as major earth quakes, stock market crashes, heavy rainfall events, etc., occur more frequently than indicated by the normal curve, i.e. they exhibit a probability distribution with a *fat tail*. Fat tails indicate a power-law pattern and interdependence. The 'tails' of a power-law curve—the regions to either side that correspond to large fluctuations—fall off very slowly in comparison with those of the bell curve

(Buchanan 2004). The normal distribution is therefore an inadequate model for extreme departures from the mean (Selvam 2009).

The following references are cited by Goertzel and Fashing (1981) to show that the bell curve is an empirical model without supporting theoretical basis: (i) Modern texts usually recognize that there is no theoretical justification for the use of the normal curve, but justify using it as a convenience (Cronbach 1970). (ii) The bell curve came to be generally accepted, as M. Lippmnan remarked to Poincare (Bradley 1968), because '…*the experimenters fancy that it is a theorem in mathematics and the mathematicians that it is an experimental fact*'. (iii) Karl Pearson (best known today for the invention of the product-moment correlation coefficient) used his newly developed chi-square test to check how closely a number of empirical distributions of supposedly random errors fitted the bell curve. He found that many of the distributions that had been cited in the literature as fitting the normal curve were actually significantly different from it, and concluded that 'the normal curve of error possesses no special fitness for describing errors or deviations such as arise either in observing practice or in nature' (Pearson 1900).

For well over a century, evidence had been mounting that real-world behaviour in particular, behaviour of *systems*, whether natural, social, economic, or financial does not follow normal distribution characteristics. The first statistical tests for normality were first developed in the 1870s by German statistician Wilhelm Lexis who found that the only series which closely matched the Gaussian distribution was birth rates. The data sets used by Pierce (Wilson and Hilferty 1929) were reexamined in 1929 by E.B. Wilson and M.M. Hilferty using formal statistical techniques which ruled out normality. The following period saw increased evidence for non-normality in real-world settings and in its place an alternative distribution, namely the power-law distribution was shown to be exhibited by real-world systems in all fields of science (Haldane 2012).

2.4 Power Laws—History

Fractals conform to power laws. A power law is a relationship in which one quantity A is proportional to another B taken to some power n; that is, $A \sim B^n$ (Buchanan 2004). Power-law distributions are found in a vast variety of systems with apparently different characteristics and have a long history of being recorded in different fields of investigation (Montroll and Shlesinger 1982, 1983, 1984). One of the oldest scaling laws in geophysics is the Omori law (Omori 1894). This law describes the temporal distribution of the number of aftershocks, which occur after a larger earthquake (i.e. mainshock) by a scaling relationship. Richardson (1960) came close to the concept of fractals when he noted that the estimated length of an irregular coastline scales with the length of the measuring unit. Richardson (1926) measured the increasing span of plumes of smoke from chimneys subjected to fluctuating atmospheric wind fields. From his observations he speculated that the turbulent air speed, which was known to be non-differentiable, could be characterized by a Weierstrass function, i.e. a self-similar fractal. He observed that the span of the plume increased as t^{β} with $\beta \geq 3$, a value inconsistent with molecular diffusion for which $\beta = 1$. The eddy cascade model of turbulence invented by Kolmogorov (1941) was, in fact, a dynamic fractal so that turbulence has no characteristic space or time scale (West 2014). References to earliest known work on power-law relationships (Mitzenmacher 2003; Andriani and McKelvey 2007; Baek et al. 2011) are summarized as follows. Pareto (1896, 1897) first noticed power laws and fat tails in economics. Cities follow a power law when ranked by population (Auerbach 1913). Dynamics of earthquakes follow power law (Gutenberg and Richter 1944, 1956) and Zipf (1949) found that a power law applies to word frequencies [Estoup (1916) had earlier found a similar relationship]. Mandelbrot (1963) rediscovered them in the twentieth century, spurring a small wave of interest in finance (Fama 1965; Montroll and Shlesinger 1984). However, the rise of the 'standard' model (Gaussian) of efficient markets sent power-law models into obscurity. This lasted until the 1990s, when the occurrence of catastrophic events, such as the 1987 and 1998 financial crashes, that were difficult to explain with the 'standard' models (Bouchaud et al. 1998), rekindled the fractal model (Mandelbrot and Hudson 2004).

A power-law world is dominated by extreme events ignored in a Gaussian world. In fact, the fat tails of power-law distributions make large extreme events orders of magnitude more likely. Theories explaining power laws are also scale free. This is to say, the same explanation (theory) applies at all levels of analysis (Andriani and McKelvey 2007).

Cumulative probability distributions which follow power-law distributions are sometimes known as Pareto distribution or Zipf's law after the two early researchers Pareto (1897) and Zipf (1949) who first investigated these distributions. Power law for cumulative probability distribution implies power law for probability P also and therefore Zipf's law or the Pareto distribution apply to power-law distributions which give the probability of occurrence of particular value of some quantity as proportional to the inverse power of that value and is found in all branches of science and other fields of study. It may be of historical interest to note that the cumulative probability distribution $P(x)$ versus x was plotted with $P(x)$ on the vertical axis by Zipf while Pareto plotted $P(x)$ on the horizontal axis. The origin of power-law behaviour has been a field of intensive study in the scientific community for more than a century (Newman 2005).

2.5 Power-Law Distributions and Complex Systems

Mathematically, the power-law distribution is written as: $P(X > x) \sim x^{-\beta}$ or $f(x) = \mathrm{K}x^{-\beta}$.

In the above expression, the exponent β is a constant for the range of x values satisfying the power-law relation and K is the constant of proportionality. Power laws appear in many different contexts. The most common are that $f(x)$ describes a distribution of random variables or the autocorrelation function of a random process. The function $f(x)$ represents the probability P that random variable X exceeds some level x is proportional to $1/x^{\beta}$, i.e. the probability of large events decays polynomially with their size. For a Gaussian distribution, the probability of large events decays *exponentially* with their size so that the probability of large events decreases much more rapidly than in the case of power laws. A Gaussian distribution predicts near-zero probability for extreme events, while the power law predicts nonzero probability as happening in real-world extreme events such as earthquakes, catastrophic floods, etc. Normal distribution gives a complete description with only the mean and variance. Power laws exhibit fat tails and cannot be described completely with a single mean and variance. Power-law-distributed data have a well-defined mean only when β lies above unity and their variance only exists when β exceeds two. Therefore, Laplace's central limit theorem may not apply to power-law-distributed variables. The normal statistical probabilities of observed variables cannot be given with reference to standard deviation and departure from the mean. Physical laws generate *system*-wide nonlinear behaviour and generate the observed power laws or fat tail distributions. As compared to Gaussian distribution, power laws give a realistic higher probability for extreme events. Power laws are a signature of interdependence of component parts in real-world complex systems. Gaussian distribution is based on central limit theorem with assumption of independence of observations and therefore not applicable to real-world complex systems (Haldane 2012). Fractals, $1/f$ noise, and Zipf's laws ubiquitous to real-world systems represent three signatures of complex systems and are associated with scaling laws. Both $1/f$ spectra and Zipf's law can be converted into a self-similar hierarchy. The mathematical laws governing this hierarchical structure when identified can provide us with a unified view of looking at complexity and complex systems (Chen 2012). A complex system consists of a large number of nonidentical components with their own complex internal structure. Local and nonlinear interactions of the internal constituents result in new emergent dynamics most frequently exhibited as the heavy tailed power-law distribution for the variables characterizing the dynamics of the complex system. The probability of observing an extremely large value is more likely in a heavy tailed distribution than in an exponential distribution such as the statistical normal distribution (Markovic and Gros 2013).

2.6 Power Laws, Scale Invariance and Self-similarity

Kavvas et al. (2015) discuss the physics of scaling and self-similarity in hydrologic dynamics, hydrodynamics, and climate as follows. Since the work of Buckingham (1914) about a century ago on dimensional analysis, the role of scales in geophysical processes has been studied by many scientists and engineers. Geophysical processes exhibit self-similarity and scale invariance over a wide range of time and space scales. Identification of the physical mechanism underlying the observed ordered cooperative existence of fluctuations on all space-time scales will help formulate simple scale-free models for simulation and prediction of geophysical processes.

Power laws are signatures of scale-free or scale-invariant phenomena, the functional form being independent of the magnitude of the scale (space or time) as explained in the following. In power laws, namely $f(x) = Kx^{-\beta}$ let the variable x undergo a scale transformation of the form $x \rightarrow cx$ where $c > 0$ is any constant. The function $f(x)$ is now transformed as $f(x) \rightarrow Kc^{-\beta}x^{-\beta} = c^{-\beta}f(x) = K_1 f(x)$, where K_1 is a constant equal to $c^{-\beta}$. The functional form of the power-law relation remains the same for scale change of the independent variable x, i.e. from x to cx. Thus power laws are a necessary and sufficient condition for scale-free behaviour (Amaral et al. 1997). Power laws imply that the physical mechanism required to generate system-wide scale-free behaviour is the same for all the scales from smallest to the largest. Power laws result from a linear relationship between logarithms such as $\log f(x) = -\beta \log x + \log K$ as seen from power law relation, namely $f(x) \sim x^{\beta}$ which is the basic definition of a power law (Farmer and Geanakoplos 2008). Scale-free behaviour signifies that the shape of the distribution $f(x)$ is unchanged, except for an overall multiplicative constant, i.e. functional form for statistical properties of the system is independent of scale size. The statistical properties of the system for any two scale sizes are related to each other by the ratio of the two scales and not on the scales themselves (Pruessner 2004).

2.7 Power Laws, Self-similarity, and Fractals

Mandelbrot (1977, 1983, 1997) from the late 1960s onwards was the first to notice and study the importance and ubiquity of scale-free behaviour in real-world systems, e.g. the coastline which exhibits a zigzag pattern on all scales of measurement. Mandelbrot gave the name 'fractals' to such self-similar structures, where the geometry of the larger structure and the enclosed smaller structures is the same. Fractals are non-differentiable geometric objects that satisfy the power-law scaling relation $f(x) = Kx^{-\beta}$ when β is not equal to an integer. Fractals are ubiquitous in nature, e.g. branches of plants, trees, river networks and tributaries, clouds, earthquakes, etc. Fractal objects cannot be described by traditional Euclidean geometry. It is found that a power-law relation exists between size and measurement resolution for a fractal object. The length of a coastline depends on the yardstick length, increasing when measured accurately with a smaller yardstick; the length varies as a power-law function of measurement resolution. Similarly, the volume of a cloud depends on the water concentration threshold and varies as a power law, inversely with this threshold. For earthquakes, floods, stock market indices, etc., the probability of large events greater than a given size, decreases with event size according to a power law. Mandelbrot brought to the world view for the first time that objects in nature are non-Euclidean and can be described by non-differentiable geometry only (Farmer and Geanakoplos 2008).

The apparently irregular space-time fluctuations exhibited by dynamical systems in nature such as rainfall, heartbeat intervals, stock market prices, etc., are associated with basic bifurcation or branching geometry of wrinkles or folds on all scales is associated with the symmetry of self-similarity under scales transformation or just self-similarity (Liu 1992).

2.8 Power Laws, 1/*f* Noise, and Long-Term Memory

For random processes that are correlated in space (or time) the autocorrelation function and the power spectrum (variance versus frequency *f*) decays as a power law signifying long-range dependence in space and time. Power spectrum exhibiting such inverse power-law form, namely $1/f^{\alpha}$, where f is the frequency and α, a constant, is also called 1/*f* noise. 1/*f* noise is exhibited by a wide range of systems in nature and signifies long-range correlations of fluctuations in space and time, identified as long-term memory or persistence. The physics of 1/*f* noise is not yet identified and is an area of intensive research in all disciplines in particular for predictability of future extreme events such as floods, earthquakes, business market crashes, etc. Power-law scaling applies strictly for the region, where the value of α remains a constant with a lower and upper level cutoff (Farmer and Geanakoplos 2008).

The association of fractals and scaling with long-range correlations or memory was brought to the attention of scientific community by the pioneering work of B. Mandelbrot in the early 1960s (Mandelbrot 1965a, b, c, 1983; Mandelbrot and Van Ness 1968; Mandelbrot and Wallis 1968, 1969a, b) to explain the Hurst effect (Hurst 1951) in hydrology related to water levels in the Nile river which exhibited anomalously fast growth of the rescaled range of the time series. Long memory signifies correlation between the present and all points in the past. A finite variance stationary process has long memory or persistence if its autocorrelation function (ACF) decays, as power law indicates fluctuations occur at all frequencies, in particular for low frequencies or large time periods. Because of this presence of low frequencies, one might observe long periods of 'high' values followed by long periods of 'low' values. This is in contrast to many standard (stationary) random processes, where the effect of each data point decays so fast that it rapidly becomes indistinguishable from noise. The long memory processes has important applications in many fields such as computer network traffic, econometrics, astrophysics, and geophysics (see e.g. Doukhan et al. 2003). The existence, or not, of long memory has special importance for predictability in weather systems particularly anthropogenic warming trend related climate change (Franzke 2012). According to the Weiner–Khintchine theorem, the spectral density function can be expressed as the Fourier transform of the autocorrelation function for a random process (Ghil and Robertson 2002). Power spectra of meteorological parameters for a wide range of space and time scales show 1/*f* noise or flicker noise, namely inverse power law form $f^{-\alpha}$, where f is the frequency and α the exponent for the variance spectrum indicating long-range correlations or memory for fluctuations in time (Lovejoy and Mandelbrot 1985; Lovejoy and Schertzer 1986a, b; Yano and Nishi 1989; Fraedrich and Larnder 1993; Fraedrich and Blender 2003; Blender and Fraedrich 2003). Traditional meteorological concepts assume characteristic scales for meteorological phenomena while observed power-law spectra of meteorological parameters show lack of characteristic scales in both space and time. Fractals, power laws and long-range correlations are signatures of nonlinear dynamics in atmospheric flows and are an intensive field of study since the 1980s (Yano et al. 2004; Bove et al. 2006; Samorodnitsky 2006, 2007; Graves et al. 2014).

2.9 Power Laws, Phase Transitions, and Critical Phenomena

Power-law distributions associated with phase transitions in critical phenomena are a well-established field in equilibrium statistical physics. The first microscopically based understanding of phase transitions is due to van der Waals, who in 1873 presented a primitive theory of the gas–liquid transition. Matter exists in three states, namely solid, liquid, and gas. The phase of a macroscopic substance is determined by a few macroscopic parameters such as the temperature and the pressure. A change in parameters such as the temperature across a phase boundary causes a sudden change in the phase of the substance. For example a solid phase changes into a liquid phase at the melting temperature. This is a phase transition. The density of the fluid has a jump at the liquid–gas phase transition (evaporation). This is the best example of a first-order phase transition, where all the physical quantities characterizing the material undergo a sudden change. Generally, the two phases are quite different at first-order transitions, and thus it takes a finite amount of energy to convert the substance from one phase to the other. This is the latent heat. It is interesting to note that the density jump at the liquid–gas transition decreases at higher pressures and temperatures. The first-order line ends at the liquid–gas critical point. The approach to this point is a second-order phase transition. In its vicinity, the fluid cannot seem to ‘decide’ what to become: a liquid or a gas. Large density fluctuations emerge leading to a ‘milky’ appearance of the fluid: the critical opalescence. Critical phenomena near continuous phase transitions at the critical points are typically observed on the scale of wavelengths of visible light (4×10^{-7} to 7×10^{-7} m) which is about three orders of magnitude larger than the random molecular scale fluctuations (H_2O molecular diameter about 2.75×10^{-10} m) of the liquid. Near the critical point, the invisible molecular scale random fluctuations form organized larger scale coherent structures visible as milky white critical opalescence.

Critical points at continuous phase transitions are one of the most interesting and important topics in equilibrium statistical physics. A phase boundary sometimes disappears at a critical point, where the two phases become indistinguishable and the substance shows anomalous behaviour, namely, the atomic/molecular scale random fluctuations organize to form macroscale coherent structures exhibiting fluctuations on all length scales. These scale-free or scale invariant fluctuations can be quantified by a power-law function with well-defined exponent called the critical exponent. All the thermodynamic properties of the system at critical point are due to the so-called scale invariance of critical phases. It is noteworthy that for a large class of very different experimental systems, the values of critical exponents are the same, independent of specific details of interactions depending only on very general properties, such as dimensionality. Near the critical point, the collective organization of individual random atomic/molecular scale fluctuations organizes to form coherent structures of all size scales observed in the scale-free fluctuations of thermodynamic parameters. Systems sharing the same values of critical exponents are said to belong to the same universality class. Universality is a well-established central concept of equilibrium physics. Systems far from equilibrium also exhibit such universality classes. However the physics of such nonequilibrium universality is not yet identified. Popkov et al. (2015) show that the two best-known examples of nonequilibrium universality classes, the diffusive and Kardar–Parisi–Zhang classes, are only part of an infinite discrete family. The members of this family are identified by their dynamical exponent, which can be expressed by a Kepler ratio of Fibonacci numbers which progressively approach the golden ratio equal to $(1 + \sqrt{5})/2$ (≈ 1.618).

It is to be noted that in equilibrium systems, the thermodynamic parameters have to be tuned precisely to observe the power-law distribution of scale-free fluctuations typically observed on the scale of wavelengths of visible light (4×10^{-7} to 7×10^{-7} m). However, spontaneous occurrence of power-law distributions has been reported in nonequilibrium systems such as turbulent atmospheric flows where coherent large-scale organization of scale-free fluctuations is observed in the size scale of up to thousands of kilometres. Peters and Neelin (2006) report similar scale-free phenomena for atmospheric precipitation on scales of tens of kilometres.

There are a number of problems in science which have as a common characteristic that complex microscopic behaviour underlies macroscopic effects. In simple cases, the microscopic fluctuations average out when larger scales are considered and the averaged quantities satisfy classical continuum equations. Hydrodynamics is a standard example of this, where atomic fluctuations average out and the classical hydrodynamic equations emerge. Unfortunately, there is a much more difficult class of problems, where fluctuations persist out to macroscopic wavelengths, and fluctuations on all intermediate length scales are important too. In fully developed turbulence in the atmosphere, global air circulation becomes unstable, leading to eddies on a scale of thousands of miles. Chaotic motions on all length scales down to millimetres are excited. Theorists have difficulties with these problems because they involve very many coupled degrees of freedom (Wilson 1982).

Mandelbrot had shown that fractal fluctuations underlie scale-free power-law distributions of thermodynamic parameters. Fractals are characterized by power-law frequency-size distributions indicating scale invariance, i.e. do not possess a characteristic length scale (Turcotte 1999). Fractals in nature are self-similar, i.e. scale invariant and do not possess a characteristic length scale, e.g. branching structure of trees and plants, river tributaries, clouds on all size scales, etc. The power-law distribution is the only function satisfying the scale-free criterion, i.e. power law is the only distribution that remains the same for any scale size.

Popkov et al. (2015) show that in nonequilibrium phenomena governed by NLFH (nonlinear fluctuating hydrodynamics) with n conservation laws, mode coupling theory predicts a family of dynamical universality classes with dynamical exponents given by the sequence of consecutive Kepler ratios of Fibonacci numbers which progressively approach the golden mean $(1 + \sqrt{5})/2$ ($\approx$1.618).

2.10 Power Laws and Self-organized Criticality

About 25 years ago, the concept of self-organized criticality (SOC) emerged (Bak et al. 1987), initially envisioned to explain the ubiquitous $1/f$-power spectra, which can be characterized by a power-law function $P(f) \propto f^{-1}$. The term *1/f power spectra* or *flicker noise* should actually be understood in broader terms, including power spectra with pink noise ($P(f) \propto f^{-1}$), red noise ($P(f) \propto f^{-2}$), and black noise ($P(f) \propto f^{-3}$), essentially everything except white noise ($P(f) \propto f^{0}$). While white noise represents traditional random processes with uncorrelated fluctuations, $1/f$ power spectra are a synonym for time series with nonrandom structures that exhibit long-range correlations (Aschwanden et al. 2016).

The signature of fractals, namely, inverse power-law form for power spectra of fluctuations was identified for isotropic homogeneous turbulence by Kolmogorov in the 1940s. The concept of fractals and its quantitative measure for space-time fluctuations of all scales was introduced by Mandelbrot in the late 1960s. The robust pattern of self-similar space-time fluctuations was identified by Bak, Tang and Wiesenfeld in the late 1980s as self-organized criticality (SOC), whereby the cooperative existence of fluctuations of all space-time scales maintains the dynamical equilibrium in dynamical systems.

Scaling behaviour is the hallmark of criticality (Fisher 1967), as it implies scale invariance: The statistical properties of a system on one scale are, apart from some factors, identical to the statistical properties of the system on another scale. The factors depend only on the ratio of the two scales and not on the scales themselves.

Power-law behaviour is observed in many different systems in nature signifying a common unifying principle underlying such scale-free organization of fluctuations. Bak et al. (1987, 1988) in their seminal work provided the central hypothesis that complex systems with many interacting components will spontaneously organize to give scale-free fluctuations such as that observed in a equilibrium thermodynamic system near a second-order phase transition known as critical phenomena, an established field of study for more than 150 years. Critical phenomena were discovered by Cagniard de la Tour in 1822 (Berchea et al. 2009). Such complex behaviour in nature occurs without the need for any fine tuning of parameters and therefore named self-organized criticality (SOC). The physics of the collective organization of short range microscopic fluctuations to result in large-scale fluctuations of all length scales with long-range correlations is not yet identified.

In equilibrium thermodynamics critical phenomena are associated with the notion of universality. Systems belonging to many different phase transitions are grouped into a small number of universality classes, each class having the same values of critical exponents and scaling functions. The collective behaviour of the system becomes independent of its microscopic details and simple models can describe the observed scale-free fluctuations. Physical systems exhibiting SOC are nonequilibrium systems, since there is a constant flux of matter and energy exchange with the environment. Universality classes, if any, such as in equilibrium phase transitions, have not yet been identified and also the dynamical processes governing the nonequilibrium phase transitions are not known (Markovic and Gros 2013).

The concept of self-organized criticality (SOC) by Bak et al. (1987, 1988) helped to bring together as a multidisciplinary field the results of earlier studies relating to power laws, 1/*f* noise, space-time fractals in disparate fields of research (e.g. van der Ziel 1950; Schick and Verveen 1974; Weissman 1988). Scale-free power-law distributions for dynamical processes indicate simple physical interactions on the smallest scale are carried over to the largest scales. SOC may therefore provide the theoretical framework for investigating in a scale-free multi-scale phenomena, the nature of the physical processes at the smallest scale (unscaled primary) operating globally at all larger scales in the continuum as visualized by Anderson (1972) and Wilson (1979) (Watkins et al. 2016).

2.11 Current Status of Power-Law Distributions

Power-law distributions, also known as heavy tail distributions, Pareto-like laws, or Zipf-like laws have been reported in real-world phenomena in different contexts. However, they are discussed only as statistical phenomena and how well they fit Pareto or Zipf laws without positive consensus regarding the fit (Clauset et al. 2009; Pinto et al. 2012; Stumpf and Porter 2012). Sornette and Ouillon (2012) state that many phenomena in nature exhibit power-law characteristics which can be related to many different physical mechanisms (Mitzenmacher 2003; Newman 2005; Sornette 2006). Scale-free distributions suggest the same dynamical mechanism for the growth of fluctuations from smaller to larger scales including extreme catastrophic events. A statistically sound power law even without a supporting rigorous theory still gives useful information regarding frequency of occurrence of extreme events (Stumpf and Porter 2012). Willinger et al. (2004) state that Mandelbrot (1997) has provided for the last 40 years mathematical, statistical, and data-analytic arguments that demonstrate that highly variable event sizes are in a sense just as 'normal'—or even more 'normal'—than Gaussian-type event sizes. Baek et al.

(2011) have given a summary of power-law distribution history and remarks that fat tails are a common feature of power-law distributions encountered in disparate contexts. The question remains whether 'fat tails' represent a global, ubiquitous nonsystem-specific feature.

2.12 Power-Law Relations (Bivariate) and Power-Law (Probability) Distributions

Reported power laws are of two different types: power-law probability distributions like the rainfall rates and bivariate power laws like allometric scaling (Stumpf and Porter 2012). Allometry designates the changes in relative dimensions of parts of the body that are correlated with changes in overall size. Julian Huxley and Georges Teissier coined this term in 1936 (Gayon 2000).

Power-law probability relations are associated with critical points and fractals. Fractals signify objects with complex geometrical properties resulting from the coexistence of many scales (Mandelbrot 1983). In statistical physics, power laws are associated with second-order phase transitions in the neighbourhood of critical points, where scale-free phenomena occur (Stumpf and Porter 2012). Power-law probability relations are associated with critical points and fractals. Bivariate power-law relation between two variables on the other hand is equivalent to a mathematical function characterizing the relationship (Sornette 2012).

Some historical bivariate power-law relations in science are (i) Coulomb's inverse square law of electrostatic force (first published in 1785). (ii) Kepler's laws of planetary motion (published between 1609–1619). (iii) Stefan–Boltzmann law of radiation (in 1879 on the basis of experimental measurements). (iv) allometric scaling (first outlined by Snell (1892), D'Arcy Thompson (1917) in *On Growth and Form*) in biology and botany.

These bivariate power-law relations are not statistical distributions, but are well established and accepted with strong data base support (Miramontes et al. 2012).

2.13 Allometric Scaling and Fractals

In biology, the allometric power-law relationship between body size x and metabolic performance y has been supported empirically over many orders of magnitude (from bacteria to whales) (West et al. 1997). West et al. (1997) proposed that one of the common mechanisms underlying allometric relationship is based on the unifying principle or assumption that a space-filling fractal-like branching structure (Mandelbrot 1977) is required for the network supplying the entire volume of the organism. Scaling laws arise from the interplay between physical and geometric constraints (West et al. 1997).

Some of the documented power laws appear to be universal to plants, animals and microbes, to terrestrial, marine and freshwater habitats; and to human-dominated as well as 'natural' ecosystems. The basic self-similar fractal architecture underlying the power laws indicates ordered organization of components of complex systems governed by a few basic physical, biological, and mathematical principles. Self-similarity over many orders of magnitude for geometrical structures and dynamical processes enable extrapolating between scales. Classic scaling relations have been described for river networks (Horton 1945; Gupta and Waymire 1989, 1998a, b; Peckham and Gupta 1999). Variations of flows, velocities, depths, widths, and slopes take the general form $Y \sim Q^b$, where Y is the hydraulic–geometric variable, Q is stream discharge and related to size of a basin, and b is a scaling exponent (Leopold and Miller 1956; Ibbitt et al. 1998; Brown et al. 2002).

Fractals are hierarchical branching networks. Nature has evolved fractal-like networks in living systems (animals and plants), where maximum surface area for energy exchange with minimum transport distance helps maximize both metabolic capacity and internal efficiency. Scaling laws typically reflect underlying general features and principles that are independent of detailed structure, dynamics or other specific characteristics of the system, or of the particular models used to describe it (West et al. 1997, 1999; West and Brown 2005).

2.14 Fractals and the Golden Section in Plant Growth

Phyllotaxis is a subdivision of plant morphology in Botany and refers to the study of repeated units such as leaves around a stem, scales on a pine cone or on a pineapple, florets in the head of a daisy, and seeds in a sunflower. These units self-organize to form repeated units of beautiful geometric patterns of spirals or helices closely resembling crystals. Adler et al. (1997) have given a comprehensive history of the study of phyllotaxis. In phyllotaxis, the angular position of leaves on the stem follows a precise mathematical spiral pattern such that the divergence angle of subsequent leaves divide the whole circle (360°) into regular fractions (Cummings and Strickland 1998). Longstanding observations show that these fractions belong to a series, where numerator and denominator are Fibonacci numbers, for example, 1/2, 1/3, 2/5, 3/8, 5/13, …. In the Fibonacci sequence 1, 1, 2, 3, 5, 8, 13, …, each term after the second is the sum of the two that precede it (Adler et al. 1997) and the ratio of each term to the preceding approaches the golden section (golden ratio or golden mean) τ equal to $(1 + \sqrt{5})/2$ (≈ 1.618), an irrational number. The earliest paper on the golden ratio by Ackermann (1895) attracted the interest of scientists and engineers in various fields of science and engineering (Li and Zhao 2013a, b). Leonardo Fibonacci of Pisa (1202) developed the Fibonacci series as a solution to the problem of monthly growth of rabbit population. The series 1/2, 1/3, 2/5, 3/8, 5/13, converges to the irrational limit of $1/\tau^2$ equal to 0.382…, which corresponds to the 'golden angle' equal to 137.5°, the golden section of full circle (360°). Serious study of phyllotaxis as a subject has a history of more than 300 years with observational phyllotaxis by Bonnet (1754), pioneering work on mathematical phyllotaxis by Braun (1831), Schimper (1836) and Bravais and Bravais (1837). Jean (1994) gives a complete survey of the studies in phyllotaxis. The physical mechanism creating these beautiful mathematical patterns of phyllotaxis is still not identified (King et al. 2004). In plants, the golden section is an essential feature found to occur with mathematical precision in the fractal-like branching patterns of trees, shrubs, etc., in the arrangement of petals/seeds and often the spiral arrangement of plant organs (Zeng and Wang 2009).

2.15 Turbulent Fluid Flow Structure, Fractals, and the Golden Ratio

von Karman (1930) showed that turbulent fluid flow dynamics near a wall follow the 'law of the wall', namely the average velocity of a turbulent flow at a certain point is proportional to the logarithm of the distance from that point to the 'wall', or the boundary of the fluid region. Recently, Li and Zhao (2013a, b) have shown that the power spectra for turbulent fluid flows derived by Kolmogorov (1941) and von Karman (1948) exhibit self-similar, i.e. fractal character of turbulent eddy fluctuations and can be expressed in terms of the golden ratio. Therefore, the trajectory of turbulent fluid flows near a boundary or wall follows a logarithmic spiral with the golden mean as the winding number similar to the spiralling arrangement of successive leaves in plant growth (Jean 1994). Selvam (1990) has shown that the atmospheric flow structure follows a logarithmic spiral trajectory with crossing angle equal to the golden mean.

2.16 Fractal Space-Time and the Golden Section

El Naschie (1994) refers to earlier work of Mauldin and Williams (1986) to show that a random Cantor set (fractal) with a Golden Mean Hausdorff dimension is a model for micro space-time.

Self-similar fractal space-time fluctuations result from the superimposition of eddy continuum fluctuations. A general systems theory model for fractal space-time fluctuations in turbulent fluid flows such as atmospheric flows (Selvam 1990, 2007, 2015) predicts that the golden mean is incorporated in the growth processes of eddy continuum fluctuations observed in atmospheric flows.

2.17 Power-Law (Probability) Distributions in Meteorological Parameters

Fraedrich and Larnder (1993) discuss earlier studies of power spectral analyses and identification of power-law distribution characteristics of meteorological parameters as follows. Power spectral analysis, namely the spectral variance density distribution of an observed time series, is used to locate dominant timescales which exhibit peaks in their contribution to the total variance. A famous example is the spectrum of high resolution with measurements at experimental sites (Van der Hoven 1957; Fiedler and Panofsky 1970; Vinnichenko 1970). Other examples, to mention a few, are the frequency spectra deduced from daily rawin observations at weather ships and continental stations (Hartmann 1974), wavenumber–frequency spectra of travelling and quasi-stationary disturbances (Fraedrich and Bottger 1978).

It is now recognized that self-similar fractal fluctuations are ubiquitous to dynamical systems such as atmospheric flows in nature. The atmospheric flow dynamics is characterized by a scale invariant broadband spectrum and cannot be completely defined by one or several prominent spectral peaks or modes. Scale invariance or scaling behaviour within a waveband implies long-range correlations between the fluctuations of the larger and smaller scale components. Thus, the scaling behaviour within a frequency band (and not the spectral peak) may be more useful and appropriate to characterize the timescales of atmospheric dynamics.

Earlier scale invariance in atmospheric flow structure was identified by (i) Richardson (1926) who applied the Lagrangian two-particle dispersion method to two-dimensional diffusion leading to his celebrated scaling law (ii) the Kolmogoroff 5/3-spectrum for isotropic three dimensional turbulence (iii) Charney's Ansatz (Charney 1971) for geostrophic turbulence, etc. More recently, in their analysis of temperatures, Lovejoy and Schertzer (1986b) characterize climatological regimes by scale invariance; this was extended to rainfall (Ladoy et al. 1991) with the emphasis to describe extreme variability by scaling and intermittency.

Fraedrich and Larnder (1993) analysed the scaling behaviour of rainfall both for a range of scales in time and for a given scale in intensity using the statistics of the Fourier transform and the cumulative probability distribution. The data sets used are (i) sets of long time series of daily rainfall at 13 European stations and (ii) sets of 5-min totals thus covering a wide scaling range.

Some early studies on application of self-organized criticality to atmospheric flows are given in the following. Turcotte (1999) has cited the following references for applications of the concept of self-organized criticality in atmospheric sciences: (i) Grieger (1992) has discussed applications to climate fluctuations (ii) Nagel and Raschke (1992) consider cloud formation processes (iii) Andrade et al. (1998) have analysed rainfall records for possible relation to self-organized criticality.

Yano et al. (2004) state that numerous studies show that meteorological phenomena exhibit scale invariant behaviour manifested as power laws in spectra (Lovejoy and Mandelbrot 1985; Lovejoy and Schertzer 1986a, b; Yano and Nishi

1989; Fraedrich and Larnder 1993; Fraedrich and Blender 2003; Blender and Fraedrich 2003). The observed scale invariant behaviour implies lack of characteristic scales in both time and space in direct contrast with traditional meteorological view of characteristic scales for individual phenomena.

Earlier Yano et al. (2001) found $1/f$ noise, a signature of scale invariance in tropical convective variability. The $1/f$ noise is unique among the power-law spectra in the sense that it contains equal variability for all timescales (Keshner 1982).

Blender et al. (2011) summarized observations and modelling of $1/f$-noise in weather and climate as follows. Meteorological and hydrological parameters exhibit fluctuations on all scales (Jiang et al. 2005; Yano et al. 2004; Fraedrich and Blender 2003) and in the absence of external influences may be attributed to the internal variability of weather and climate. The low frequency part of the power spectra of these time series exhibit approximate inverse power-law form $f^{-\alpha}$ with $\alpha > 0$ and signifies long-range space-time correlations or long-term memory.

Data with power spectra close to $S(f) \sim 1/f$ is denoted as $1/f$ or flicker noise. Examples for $1/f$ spectra are found in wide ranges from minutes to millennia: (i) tropical boundary layer observations (TOGA/COARE, Yano et al. 2004) (ii) the discharge of the Yangtze river in the intra-annual frequency range (Wang et al. 2008) (iii) the sea surface temperature in a region of the North Atlantic and in the southern ocean [reproduced by models (Fraedrich and Blender 2003)] (iv) reconstructed near surface temperatures up to the Milankovitch cycle (Huybers and Curry 2006).

Andrade et al. (1998) state that long-term data sets from weather stations around the world exhibit self-organized critical dynamics, namely, the distribution of droughts in semiarid regions obeys a clear-cut power law. The statistics for rain intensity, on the other hand, exhibit two distinct scaling regimes. These authors cite the following studies which show scale-free power-law distributions, a signature of SOC in meteorological parameters.

1. Analysis of historical records of the Southern Pacific sea surface temperature by Andrade et al. (1995) indicate that the El Nino may be an example of SOC in climate phenomena.
2. It is now recognized that several properties of the statistical distributions of pertinent meteorological fields (temperature, air humidity, etc.) are independent of the particular (time or length) scale at which they are observed (Lovejoy and Schertzer 1991, 2010; Vattay and Harnos 1994; Fraedrich and Larnder 1993; Ladoy et al. 1991), i.e. they obey scale-free distributions.
3. Former identifications of power-law behaviour in single precipitation events (Ladoy et al. 1991; Lovejoy and Mandelbrot 1985), and work of Fraedrich and Larnder 1993) which detects two different scaling domains in the rainfall spectrum.

2.17.1 Power-Law (Probability) Distributions in Precipitation

Arakawa (2006) remarks that despite the complexity of the processes involved, the statistics of the tropical rain rate are remarkably similar to those of critical phenomena near continuous phase transitions in other—much smaller—physical systems.

Deluca et al. (2015) state that coupling between several nonlinear mechanisms with different spatial and temporal characteristic scales gives rise to structures and correlations across long ranges in space and time in the observed characteristics of precipitation (Lovejoy 1982; Vattay and Harnos 1994; Yano et al. 2004; Bodenschatz et al. 2010). Strong statistical regularities are presented by individual rain events from diverse backgrounds (Peters et al. 2002, 2010; Peters and Neelin 2006; Neelin et al. 2008; Deluca and Corral 2014; Deluca et al. 2016) giving support to the hypothesis that atmospheric convection and precipitation may be a real-world example of self-organized criticality (SOC).

Jordan (2008) gives a summary of observational studies of region wise evidence for the operation of SOC in atmospheric convective processes as follows.

(i) Vattay and Harnos (1994) were the first to make the suggestion that SOC is present in the atmosphere. They showed that the daily average air humidity fluctuations from central Europe over the interval of a year exhibited approximate $1/f$ behaviour.

(ii) Later Peters and Christensen (2002) using rainfall statistics at a site on the German Baltic coast over an 8 month period from January to July 1999 showed that no typical size or time scale exists for rain event sizes and durations confirming signature of SOC in atmospheric processes.

(iii) A detailed analysis of tropical convective variability using Tropical Ocean Global Atmosphere Coupled Ocean Atmosphere Response Experiment (TOGA-COARE) data was carried out by Yano et al. (2003). Using time series data from 13 sites in the western Pacific, approximate $1/f$ behaviour was found to occur in atmospheric surface variables (air temperature, moisture mixing ratio and wind speed) over an interval from 1 h to 10 days.

(iv) Critical phenomena in tropical rain was first studied by Peters and Neelin (2006) using tropical rainfall data. The data sets that the study analysed were comprehensive; satellite microwave data from each major global ocean basin over a 5 year period from 2000 to 2005 at a 20 km grid resolution. Analysing the relationship between vertically integrated water vapour and precipitation, they demonstrated that above a critical value of vertically integrated water vapour precipitation is intense, short-lived, and follows a power-law relationship. Below that critical value, it is weak but more persistent.

Andrade et al. (1998), Peters et al. (2002) and Peters and Christensen (2002, 2006) were the first to study rain events as avalanches in SOC of precipitation processes. These authors defined, independently, the event size as equal to the total amount of rain collected during the duration of the event. There was evidence for power-law distributions for event sizes and for dry-spells durations over several orders of magnitude. A later study by Peters et al. (2010) covering 10 sites across different climates using rain data from optical gauges showed unambiguous power-law distributions of event sizes and dry-spell durations. Peters et al. (2010) compared rain event size distributions derived from measurements in climatically different regions and found them to be well approximated by power laws of similar exponents over broad ranges. Differences were seen in the large-scale cutoffs of the distributions. Event duration distributions suggested that the scale-free aspects are related to the absence of characteristic scales in the meteorological mesoscale.

Deluca and Corral (2014) analysed distributions of rain event sizes, rain event durations, and dry-spell durations for data obtained from a network of 20 rain gauges scattered in a region of the NW Mediterranean coast. The distributions exhibit scale invariance, a signature of SOC in precipitation observed for medium resolution rain data.

Rain fields exhibit large spatial and temporal intermittency and extreme variability such that their intensity cannot be characterized by its mean value (Bodenschatz et al. 2010). The complex phenomena of cloud and rain formation processes exhibit surprising statistical regularities such as (i) numerous geometric and radiative properties of clouds present clear scaling or multiscaling behaviour (Lovejoy 1982; Cahalan and Joseph 1989; Peters et al. 2009; Wood and Field 2011) (ii) raindrop arrival times and raindrop sizes, are well characterized by power-law distributions over several of orders of magnitude (Olsson et al. 1993; Lavergnat and Gole 2006).

Peters and Christensen (2002) show that the number density of rain events per year is inversely proportional to the released water column raised to the power 1.4. This is the rain equivalent of the Gutenberg–Richter law (Gutenberg and Richter 1944, 1956) for earthquakes. The event durations and the waiting times between events are also characterized by scaling regions, where no typical time scale exists. Scale-free power-law behaviour is found to govern the statistics of rain over a wide range of time and event size scales and rain is an excellent example of a self-organized critical process.

Processes relevant for precipitation are associated with many different characteristic time and spatial scales, see, e.g. (Bodenschatz et al. 2010). The list of these scales has a gap, however, from a few kilometres (a few minutes) to 1000 km (a few days), spanning the so-called mesoscale.

Devineni et al. (2015) present the first ever results on a global analysis of the scaling characteristics of extreme rainfall areas for durations ranging from 1 to 30 days. Their findings lead to the question as to how the climate system organizes over these scales, overcoming the substantial apparent heterogeneity in process dynamics. They suggest that power-law scaling may also apply to planetary scale phenomenon, such as frontal and monsoonal systems, and their interaction with local moisture recycling. Such features may have persistence over large areas corresponding to extreme rain and regional flood events.

Peters et al. (2002) showed that power law is followed by precipitation observations, with the coefficient α being 1.36 and 1.42 for precipitation intensity and drought length, respectively. More recently, Bove et al. (2006) found power-law distributions of precipitation intensity and drought length in 15-min resolution rain gauge data, with coefficients of 2.35 and 2.1 for intensity and drought, respectively. Their analysis of rain data collected from automatic stations (Data Collection Platforms) spread on the Italian territory shows that the rain is a self-organized critical phenomenon. They show that power laws describe the number of rain events versus size and number of droughts versus duration. Anomalous Hurst coefficients and one-over-*f* (1/*f*) noise found are consistent with the concepts of criticality and self-similarity.

Peters and Neelin (2006) analysed satellite microwave estimates of rainfall rate, *P*, water vapour, *w*, cloud liquid water and sea surface temperature (SST) from the Tropical Rainfall Measuring Mission from 2000 to 2005. Observations from the western Pacific provided initial support for their conjecture, namely, a power-law pickup of precipitation (the order parameter) above a critical value, w_c, of water vapour (the tuning parameter).

Atmospheric convection and precipitation have been hypothesized to be a real-world realization of self-organized criticality (SOC). This idea is supported by observations of avalanche-like rainfall events (Andrade et al. 1998; Peters et al. 2002) and by the nature of the transition to convection in the atmosphere (Peters and Neelin 2006; Neelin et al. 2009).

Wang and Huang (2012) analysed the long-term rain records of five meteorological stations in Henan, a central province of China for rain duration, drought duration, accumulated rain amount characterizing these rain events processes. They found that the long-term rain processes in central China exhibit the feature of self-organized criticality.

Andrade analysed long-term daily rain records of weather stations around the world with a special emphasis on the semiarid regions and found that there existed some evidences of SOC with these data (Andrade et al. 1998). Peters et al. investigated the European rain and found that it exhibits the feature of SOC (Peters and Christensen 2002; Peters et al. 2002; Pruessner and Peters 2006).

Pelino et al. (2006) show that daily precipitation is characterized by a behaviour regulated by power laws for the frequency distribution of both event intensity and drought duration. In this respect, precipitation appears to follow self-organized criticality laws, much as other geophysical phenomena such as avalanches and earthquakes.

Peters et al. (2002) analysed data at 1-min resolution showed that this power law is followed by precipitation observations for precipitation intensity and drought length.

Bove et al. (2006) found power-law distributions of precipitation intensity and drought length in 15-min resolution rain gauge data. Scale-free power-law behaviour is found to govern the statistics of rain over a wide range of time and event size scales (Peters and Christensen 2006).

Therefore observed precipitation exhibits self-organized criticality and complexity-like behaviour, at least at fine temporal scales.

Sarkar and Barat (2006) analysed monthly rainfall records of All India (total) and different regions of India for the time period 1871–2002 and found that the distributions of the rainfall intensity exhibit perfect power-law behaviour. The scaling analysis revealed two distinct scaling regions in the rainfall time series.

2.17.2 Power-Law (Probability) Distributions in Temperature

Franzke's (2012) study investigates the significance of trends of four temperature time series—Central England Temperature (CET), Stockholm, Faraday–Vernadsky, and Alert. An analysis of the four temperature time series reveals evidence of long-range dependence (LRD) and nonlinear warming trends. There is increasing evidence that surface temperatures are long-range dependent (Koscielny-Bunde et al. 1998; Eichner et al. 2003; Gil-Alana 2005; Huybers and Curry 2006; Rybski et al. 2006; Fatichi et al. 2009; Franzke 2010).

Fraedrich et al. (2009) cite the following studies for long-term memory and $1/f$ noise in climate variability. The variability of the near surface temperature shows a continuum of long-term memory. In the atmosphere, the high frequency range is assessed by observations in the tropical western Pacific. Here, convective available potential energy (CAPE) shows a $1/f$ spectrum within 1–30 days (Yano et al. 2001), while temperature, wind speed, and moisture show this spectrum within 1 h to 10 days. The observed sea surface temperature in the North Atlantic shows a $1/f$ spectrum on intra-annual timescales (Fraedrich and Blender 2003). Monetti et al. (2003) studied the persistence in sea surface temperature (SST) records at many sites in the Atlantic and Pacific oceans. They find that for all timescales, the SST fluctuations exhibit stronger correlations than atmospheric land temperature fluctuations (Monetti et al. 2003). Greenland ice cores present a proxy for temperature variability revealing scaling long-term memory on millennial timescales (Blender et al. 2006).

Fraedrich et al. (2004) state that spectra of observed ocean surface temperatures follow $1/f$ (or flicker noise) beyond one year in northern and southern midlatitudes (Fraedrich and Blender 2003; Monetti et al. 2003).

Liu et al. (2014) investigated the frequency-size distribution of three climatic factors (average daily temperature, vapour pressure, and relative humidity) for the period 1961–2011 in Yanqi County, northwest China and found that they were well approximated by power-law distribution, which suggested that climatic factor might be a manifestation of self-organized criticality.

2.17.3 Power-Law (Probability) Distributions in Quaternary Ice Volume Fluctuations

Grieger (1992) shows that the concept of self-organized criticality due to Bak et al. (1987, 1988) offers a simple and appealing possibility to explain the power-law background spectrum of the quaternary ice volume fluctuations.

2.17.4 Power-Law (Probability) Distributions in Atmospheric Pollution

Kai et al. (2013) found long-range correlation at one-year temporal scale in their analysis of long-term time series of daily average PM10 (pollution index) concentrations in Chengdu city, China. Further spectral analysis of the time series indicated $1/f$ noise behaviour. The probability distribution functions of PM10 concentrations fluctuation have a scale-invariant structure. The observed scale-invariant structure of PM10 fluctuations indicates the operation of SOC in atmospheric dynamical processes.

Liu et al. (2015) analysed the temporal fluctuations of the three pollution indices (SO_2, NO_2 and PM10) and the daily air pollution indices (APIs) of Shanghai in China. The results show that the temporal scaling behaviours in all the four series exhibit two different power laws. Their findings suggest that SO_2, NO_2, and PM10 pollution is an example of a self-organized criticality (SOC) process in the atmosphere.

2.18 General Systems Theory Model for Self-organized Criticality in Atmospheric Flows

Selvam (1990, 2007, 2014, 2015) has presented a general systems theory model for turbulent fluid flows such as atmospheric flows. The three important model predictions are given as follows:

(a) The observed fractal fluctuations of meteorological parameters result from the superimposition of an eddy continuum fluctuations.
(b) The probability distribution P of amplitudes of component eddies represents the variance distribution also and is given in Eq. (2.1) as

$$P = \tau^{-4t}, \tag{2.1}$$

where τ ($\approx$1.618) is the golden mean and t the normalized deviation (deviation/standard deviation) of the data used for the study. Popkov et al. (2015) show that in nonequilibrium phenomena governed by NLFH (nonlinear fluctuating hydrodynamics) with n conservation laws, mode coupling theory predicts a family of dynamical universality classes with dynamical exponents given by the sequence of consecutive Kepler ratios of Fibonacci numbers which progressively approach the golden mean $(1 + \sqrt{5})/2$ ($\approx$1.618).

(c) A universal (scale independent) spectrum for suspended atmospheric particulate size distribution expressed as a function of the golden mean τ ($\approx$1.618), the total number concentration and the mean volume radius (or diameter) of the particulate size spectrum. A knowledge of the mean volume radius and total number concentration is sufficient to compute the total particulate size spectrum at any location.

The model prediction at (b) implies that the additive amplitudes of eddies when squared (variance) represent the probability distribution of eddies and such a result is observed in the subatomic dynamics of quantum systems such as the electron or photon. Therefore atmospheric flows and in general turbulent fluid flows follow quantum-like dynamics.

The model predicted probability distribution P (Eq. 2.1) of fluctuation amplitudes is very close to the statistical normal distribution for values of normalized deviation t less than 2 and gives progressively larger values than the normal distribution for normalized deviation t values greater than 2. The statistical normal distribution gives near-zero probability of occurrence of extreme events ($t \gg 2$) while the model predicts appreciable probabilities as observed in practice.

The model predicted probability distribution for fractal fluctuations is shown to be the same as the Boltzmann's distribution for molecular scale energy distribution.

The general system theory predicted inverse power-law distribution for fluctuation amplitudes and variance distribution at Eq. (2.1) is a signature of self-organized criticality in atmospheric flows.

The following list of ten continuous periodogram power spectral analyses of data sets for different meteorological parameters for different time periods show that the variance spectra follow model predicted inverse power-law distribution at Eq. (2.1), a signature of self-organized criticality in atmospheric flows.

(i) Three sets each of 30 years (1871–1900, 1901–1930, 1956–1985) and one set of 25 years (1931–1955) summer monsoon rainfall time series for 29 meteorological subdivisions in the Indian region (Selvam et al. 1992).
(ii) 115 years (1871–1985) summer monsoon rainfall over the Indian region (Selvam 1993).
(iii) Sets of 50–364 daily mean atmospheric columnar total ozone content at 19 globally representative stations (Selvam and Radhamani 1994).
(iv) 28 years (1961–1988) of seasonal (September–November) mean COADS global surface (air and sea) temperature time series (Selvam and Joshi 1995).

(v) Daily values of atmospheric total ozone for five different stations were obtained from Ozone Data for the World (Dept. of Environment) 1988–91. Data sets used for the study are the following six sets of 20 to 25 daily or up to 6 days averages of atmospheric total ozone content. (1) Hobart (Australia): daily values for the period 1–25 February 1991, (2) Melbourne (Australia): 2-day mean values for the period 1 October–19 November 1990, (3) Melbourne (Australia): daily values for the period 1 February–20 February 1991, (4) Tateno (Japan): 6-day mean values for the period 1 March–30 April 1990, (5) Reykjavik (Iceland): 6-day mean values for the period 3 March–30 June 1990, (6) Varanasi (India) 4-day mean values for the period 1 March–19 May 1990 (Selvam and Radhamani 1995).

(vi) Two 50-years (1871–1920 and 1936–1985) of summer monsoon rainfall over the Indian region and one 84-years set (1893–1976) of winter half-year rainfall over England and Wales (Selvam et al. 1995).

(vii) Annual and seasonal mean global surface pressure time series for the 25 years 1964–1988 obtained from the Comprehensive Ocean Atmosphere Data Set (COADS) (Selvam et al. 1996).

(viii) The following data sets of climatological parameters from disparate climatic regimes: (1) Annual and Seasonal means of rainfall: Subdivision wise and total Indian region, 1871–1994 for 124 years, (2) Annual rainfall, England and Wales, 1766–1980 for 215 years, (3) Southern Oscillation Index (SOI): Seasonal, Tahiti−Darwin, 1852–1984 for 133 years, (4) Surface Temperature: Annual and Seasonal, Arctic, 1957–1981 for 25 years (5) Surface Temperature: Annual and Seasonal, Antarctic, 1957–1983 for 27 years (Selvam and Fadnavis 1998).

(ix) Two-day mean 00 GMT upper air (850, 500, and 200 mb) temperature data of TOGA (Tropical Ocean Global Atmosphere) temperature time series for latitude belts from 50°N to 50°S at 5° latitude and longitude intervals for the four seasons MAM, JJA, SON, DFJ respectively for the 5-year period 1986–1990 (Joshi and Selvam 1999).

(x) Global-gridded time series data sets of monthly mean temperatures for the period 1880–2007/2008 (Selvam 2011a).

General systems theory model prediction at (b) above that amplitudes of fractals fluctuations also follows the same inverse power-law form τ^{-4t} (Eq. 2.1) as the variance spectrum, where τ ($\approx$1.618) is the golden mean and t the normalized deviation of the data used for the study is seen in the following study. Historic month-wise temperature (maximum and minimum) and total rainfall for the four stations Oxford, Armagh, Durham and Stornoway in the UK region, for data periods ranging from 92 years to 160 years show that the fluctuations of temperature and rainfall follow the inverse power-law form τ^{-4t}, where τ is the golden mean ($\approx$1.618) and t is the normalized standard deviation equal to mean divided by the standard deviation of the data set (Selvam 2014).

The general systems theory model prediction at (c) above for fractal space-time fluctuations predicts a universal (scale independent) spectrum for suspended atmospheric particulate size distribution expressed as a function of the golden mean τ ($\approx$1.618), the total number concentration and the mean volume radius (or diameter) of the particulate size spectrum (Selvam 2015). A knowledge of the mean volume radius and total number concentration is sufficient to compute the total particulate size spectrum at any location. Model predicted spectrum is in agreement with the following four experimentally determined data sets:

(I) CIRPAS mission TARFOX_WALLOPS_SMPS aerosol size distributions (2) CIRPAS mission ARM-IOP (Ponca City, OK) aerosol size distributions (3) SAFARI 2000 CV-580 (CARG Aerosol and Cloud Data) cloud drop size distributions and (4) TWP-ICE (Darwin, Australia) rain drop size distributions (Selvam 2012a).

(II) Observations for atmospheric aerosol size distribution using VOCALS 2008 PCASP data (Selvam 2012b).

(III) The total averaged radius size spectra for the AERONET (aerosol inversions) stations Davos and Mauna Loa for the year 2010 and Izana for the year 2009 (Selvam 2013).

(IV) The following two experimentally determined atmospheric aerosol data sets, (1) SAFARI 2000 CV-580 Aerosol Data, Dry Season 2000 (CARG) (2) World Data Centre Aerosols data sets for the three stations Ny Ålesund, Pallas and Hohenpeissenberg (Selvam 2011b).

References

Abbott, D.: Overview: unsolved problems of noise and fluctuations. Chaos Interdisc. J. Nonlinear Sci. **11**(3), 526–538 (2001)

Ackermann, E.C.: The Golden Section. Am. Math. Mon. **2**(9–10), 260–264 (1895)

Adler, I., Barabe, D., Jean, R.V.: A history of the study of phyllotaxis. Ann. Bot. **80**, 231–244 (1997)

Amaral, L.A.N., Buldyrev, S.V., Havlin, S., Leschhorn, H., Maass, P., Salinger, M.A., Stanley, H. E., Stanley, M.H.R.: Scaling behaviour in economics: I. empirical results for company growth. Phys. A **244**(1–4), 1–24 (1997)

Anderson, P.W.: More is different. Sci. New Ser. **177**(4047), 393–396 (1972)

Andrade Jr., J.S., Wainer, I., Filho, J.M., Moreira, J.E.: Self-organized criticality in the El Niño southern oscillation. Phys. A **215**(3), 331–338 (1995)

Andrade, R.F.S., Schellnhuber, H.J., Claussen, M.: Analysis of rainfall records: possible relation to self-organized criticality. Phys. A (Stat. Mech. Appl.) **254**(3–4), 557–568 (1998)

Andriani, P., McKelvey, B.: Beyond Gaussian averages: redirecting management research toward extreme events and power laws. J. Int. Bus. Stud. **38**, 1212–1230 (2007)

Arakawa, A.: Atmospheric physics: scaling tropical rain. News Views Nat. Phys. **2**, 373–374 (2006)

Aschwanden, M.J., Crosby, N.B., Dimitropoulou, M., Georgoulis, M.K., Hergarten, S., McAteer, J., Milovanov, A.V., Mineshige, S., Morales, L., et al.: 25 years of self-organized criticality: solar and astrophysics. Space Sci. Rev. **198**(1), 47–166 (2016)

Auerbach, F.: Das Gesetz Der Bevolkerungskoncentration. Petermanns Geogr. Mitt. **59**, 74–76 (1913)

Baek, S.K., Bernhardsson, S., Minnhagen, P.: Zipf's law unzipped. New J. Phys. **13**, 043004, 21 (2011). http://arxiv.org/pdf/1104.1789v1.pdf

Bak, P., Tang, C., Wiesenfeld, K.: Self-organized criticality: an explanation of 1/*f* noise. Phys. Rev. Lett. **59**, 381–384 (1987)

Bak, P., Tang, C., Wiesenfeld, K.: Self-organized criticality. Phys. Rev. A **38**, 364–374 (1988)

Berche, B., Malte, H., Kenna, R. Critical phenomena: 150 years since Cagniard de la Tour. J. Phys. Stud. **13**(3), 3001, 4 (2009). arXiv:0905.1886v1[physics.hist-ph]

Blender, R., Fraedrich, K.: Long time memory in global warming simulations. Geophys. Res. Lett. **30**, 1769–1772 (2003)

Blender, R., Fraedrich, K., Hunt, B. Millennial climate variability: GCM-simulation and Greenland ice cores. Geophys. Res. Lett. **33**, L04710, 4 (2006)

Blender, R., Zhu, X., Fraedrich, K.: Observation and modelling of 1/*f* noise in weather and climate. Adv. Sci. Res. **6**, 137–140 (2011)

Bodenschatz, E., Malinowski, S.P., Shaw, R.A., Stratmann, F.: Can we understand clouds without turbulence? Science **327**(5968), 970–971 (2010)

Bonnet, C.: Recherches Sur L'usage Des Feuilles Dans Les Plantes. E. Luzac fils, Gottingen and Leyden (1754)

Bouchaud, J.P., Sornette, D., Walter, C., Aguilar, J.P.: Taming large events: optimal portfolio theory for strongly fluctuating assets. Int. J. Theor. Appl. Finan. **1**(1), 25–41 (1998)

Bove, R., Pelino, V., De Leonibus, L.: Complexity in rainfall phenomena. Commun. Nonlinear Sci. Numer. Simul. **11**(6), 678–684 (2006)

Bradley, J.V.: Distribution-Free Statistical Tests. Englewood Cliffs, Prentice-Hall, N.J. (1968)

Braun, A.: Vergleichende Untersuchung über die Ordnung der Schuppen an den Tannenpflanzen als Einleitung zur Untersuchung der Blattstellung überhaupt. Nova Acta Ph Med. Acad. Cesar Leop Carolina Nat Curiosorum **15**, 195–402 (1831)

Bravais, L., Bravais, A.: Essai sur la disposition des feuilles curvisériées. Ann. Sci. Nat. Bot. Biol. Vég. **7**, 42–110, 193–221, 91–348; **8**, 11–42 (1837)

Brown, J.H., Gupta, V.K., Li, B.-L., Milne, B.T., Restrepo, C., West, G.B.: The fractal nature of nature: power laws, ecological complexity and biodiversity. Phil. Trans. R. Soc. Lond. B **357**, 619–626 (2002)

Buchanan, M.: Power laws and the new science of complexity management. Strategy Bus. Issue **34**, 70–79 (2004)

Buckingham, E.: On physically similar systems—illustrations of the use of dimensional equations. Phys. Rev. **4**, 345–376 (1914)

Cahalan, R., Joseph, J.: Fractal statistics of cloud fields. Mon. Wea. Rev. **117**, 261–272 (1989)

Charney, J.G.: Geostrophic turbulence. J. Atmos. Sci. **28**, 1087–1095 (1971)

Chen, Y.: Zipf's law, 1/*f* noise, and fractal hierarchy. Chaos Solitons Fractals **45**, 63–73 (2012)

Clauset, A., Shalizi, C.R., Newman, M. E.J.: Power-law distributions in empirical data. SIAM Rev. **51**(4), 661–703. (2009) arXiv:0706.1062v2[physics.data-an]. 2 Feb 2009

Cronbach, L.: Essentials of Psychological Testing. Harper and Row, New York (1970)

Cummings, F.W., Strickland, J.C.: A model of phyllotaxis. J. Theor. Biol. **192**, 531–544 (1998)

Deluca, A., Corral, A.: Scale invariant events and dry spells for medium resolution local rain data. Nonlin. Process. Geophys. **21**, 555–567. (2014) arXiv:1212.5533v1[physics.ao-ph]

Deluca, A., Moloney, N. R., Corral, A.: Data-driven prediction of thresholded time series of rainfall and SOC models. Phys. Rev. E **91**(5), 052808. (2015) arXiv:1411.2256[physics.data-an]

Deluca, A., Puig, P., Corral, A.: Testing universality in critical exponents: the case of rainfall. Phys. Rev. E **93**, 042301. (2016) arXiv:1508.06516v1[physics.data-an]

Devineni, N., Lall, U., Xi, C., Ward, P.: Scaling of extreme rainfall areas at a planetary scale. Chaos **25**, 075407 (2015)

Doukhan, P., Oppenheim, G., Taqqu, M.S. (eds.): Theory and Applications of Long-Range Dependence. Birkhauser, Boston, MA (2003)

Eichner, J.F., Koscielny-Bunde, E., Bunde, A., Havlin, S., Schellnhuber, H.-J.: Power-law persistence and trends in the atmosphere: a detailed study of long temperature records. Phys. Rev. E **68**, 046133 (2003)

Einstein, A .: "Über die von der molekularkinetischen Theorie der Wärme geforderte Bewegung von in ruhenden Flüssigkeiten suspendierten Teilchen" (PDF). "Investigations on the theory of Brownian Movement". Translated by A.D Cowper. *Annalen der Physik* **17**(8), 549–560 (1905)

El Naschie, M.S.: Is quantum space a random cantor set with a golden mean dimension at the core? Chaos Solitons Fractals **4**(2), 177–179 (1994)

Estoup, J.B.: Gammes Stenographiques. Institut Stenographique de France, Paris (1916)

Fama, F.: The behavior of stock-market prices. J. Bus. **38**, 34–105 (1965)

Farmer, J.D. Geanakoplos, J.: Power laws in economics and elsewhere. (2008) www.santafe.edu/~jdf/papers/powerlaw3.pdf

Fatichi, S., Barbosa, S.M., Caporali, E., Silva, M.E.: Deterministic versus stochastic trends: Detection and challenges. J. Geophys. Res. **114**, D18121 (2009)

Fiedler, F., Panofsky, H.A.: Atmospheric scales and spectral gaps. Bull. Amer. Meteor. Soc. **51**, 1114–1119 (1970)

Fisher, M.E.: The theory of equilibrium critical phenomena. Rep. Prog. Phys. **30**(2), 615–730 (1967)

Fraedrich, K., Blender, R.: Scaling of atmosphere and ocean temperature correlations in observations and climate models. Phys. Rev. Lett. **90**, 108501, 4 (2003)

Fraedrich, K., Bottger, H.: A wavenumber frequency analysis of the 500 mb geopotential at 50°N. J. Atmos. Sci. **35**, 745–750 (1978)

Fraedrich, K., Larnder, C.: Scaling regimes of composite rainfall time series. Tellus **45A**, 289–298 (1993)

Fraedrich, K., Blender, R., Zhu, X.: Continuum climate variability: long-term memory, scaling, and 1/*f*-noise. Int. J. Mod. Phys. B **23**(28 and 29), 5403–5416 (2009)

Fraedrich, K., Luksch, U., Blender, R.: 1/*f* model for long-time memory of the ocean surface temperature. Phys. Rev. E **70**, 037301, 4 (2004)

Franzke, C.: Long-range dependence and climate noise characteristics of antarctic temperature data. J. Clim. **23**, 6074–6081 (2010)

Franzke, C.: Nonlinear trends, long-range dependence, and climate noise properties of surface temperature. J. Clim. **25**, 4172–4183 (2012)

Gayon, J.: History of the concept of allometry. Amer. Zool. **40**(5), 748–758 (2000)

Ghil, M., Robertson, A.W.: "Waves" vs. "particles" in the atmosphere's phase space: a pathway to long-range forecasting? Colloquium PNAS **99**(suppl. 1), 2493–2500. (2002) Arthur M. Sackler Colloquium of the National Academy of Sciences, "*Self-Organized Complexity in the Physical, Biological, and Social Sciences*," held March 23–24, 2001, at the Arnold and Mabel Beckman Center of the National Academies of Science and Engineering in Irvine, CA

Gil-Alana, L.A.: Statistical modeling of the temperatures in the Northern Hemisphere using fractional integration techniques. J. Clim. **18**, 5357–5369 (2005)

Goertzel, T., Fashing, J.: The myth of the normal curve: a theoretical critique and examination of its role in teaching and research. Humanity Soc. **5**, 14–31 (1981); reprinted in Readings in Humanist Sociology, General Hall, 1986. http://crab.rutgers.edu/~goertzel/normalcurve.htm 4/29/2007

Graves, T., Gramacy, R.B., Franzkez, C., Watkins, N.W.: Effcient Bayesian inference for long memory processes. Nonlin. Proc. Geophys. Discuss. **2**, 573–618. (2014) arXiv:1403.2940v1 [stat.ME]

Greene, W.H.: Econometric Analysis, 5th edn. Prentice-Hall, Englewood Cliffs, NJ (2002)

Grieger, B.: Quaternary climatic fluctuations as a consequence of self-organized criticality. Phys. A **191**, 51–56 (1992)

Gupta, V.K., Waymire, E.: Spatial variability and scale invariance in hydrologic regionalization. In Sposito, G. (ed.) Scale Dependence and Scale Invariance in Hydrology, pp. 88–135. Cambridge University Press (1998a)

Gupta, V.K., Waymire, E.: Some mathematical aspects of rainfall, landforms and floods. In Barndorff-Nielsen, O.E., Gupta, V.K., Perez-Abreu, V., Waymire, E.C. (eds.) Advanced Series in Statistical Sciences and Applied Probability, vol. 7, pp. 129–172. World Scientific, Singapore (1998b)

Gupta, V.K., Waymire, E.: Statistical self-similarity in river networks parameterized by elevation. Water Resour. Res. **25**, 463–467 (1989)

Gutenberg, B., Richter, C.F.: Frequency of earthquakes in California. Bull. Seismol. Soc. Am. **34**, 185–188 (1944)

Gutenberg, B., Richter, C.F.: Magnitude and energy of earthquakes. Ann. Geofis. **9**, 1–15 (1956)

Haldane, A.G.: Tails of the unexpected. Given at "The Credit Crisis Five Years On: Unpacking the Crisis", conference held at the University of Edinburgh Business School (2012). 8–9 June 8 June 2012. All speeches are available online at www.bankofengland.co.uk/publications/Pages/speeches/default.aspx

Hartmann, D.: Time spectral analysis of midlatitude disturbances. Mon. Wea. Rev. **102**, 348–362; Corrigendum: 541–542 (1974)

Horton, R.E.: Erosional development of streams and their drainage basins; hydrophysical approach to quantitative morphology. Bull. Geol. Soc. Am. **56**, 275–370 (1945)

Hurst, H.E.: Long-term storage capacity of reservoirs. Trans. Am. Soc. Civ. Eng. **116**, 770–808. With discussion (1951)

Huybers, P., Curry, W.: Links between annual, Milankovitch and continuum temperature variability. Nature **441**, 329–332 (2006)

Ibbitt, R.P., McKercher, A.I., Duncaeri, M.J.: Taieri River data to test channel network and river basin heterogeneity concepts. Water Resour. Res. **34**, 2085–2088 (1998)

Jean, R.V.: Phyllotaxis: A systemic study in plant morphogenesis. Cambridge University Press, Cambridge and New York (1994)

Jiang, T., Zhang, Q., Blender, R., Fraedrich, K.: Yangtze delta floods and droughts of the last millennium: Abrupt changes and long term memory. Theor. Appl. Climatol. **82**, 131–141 (2005)

Jordan, T.: Does self-organised criticality occur in the tropical convective system?. This dissertation is submitted to the Department of Mathematics and the Department of Meteorology in partial fulfilment of the requirements for the degree of Master of Science, University of Reading, School of Mathematics, Meteorology and Physics (2008)

Joshi, R.R., Selvam, A.M.: Identification of self-organized criticality in atmospheric low frequency variability. Fractals **7**(4), 421–425 (1999)

Kai, S., Chun-Qiong, L., Si-Chuan, L.: Self-organized criticality: emergent complex behavior in PM10 pollution. Int. J. Atmos. Sci. Hindawi Publishing Corporation, Article 419694, 7 (2013)

Kavvas, M.L., Govindaraju, R.S., Lall, U.: Introduction to the focus issue: physics of scaling and self-similarity in hydrologic dynamics, hydrodynamics, and climate. Chaos **25**, 075201, 2 (2015)

Keshner, M.S.: *1/f* noise. Proc. IEEE **70**, 212–218 (1982)

King, S., Beck, F., Lüttge, U.: On the mystery of the golden angle in phyllotaxis. Plant Cell Environ. **27**, 685–695 (2004)

Kolmogorov, A.N.: The local structure of turbulence in incompressible viscous fluid for very large Reynolds numbers. C. R. Acad. Sci. URSS **30**, 301–305 (1941)

Koscielny-Bunde, E., Bunde, A., Havlin, S., Roman, H.E., Goldreich, Y., Schellnhuber, H.-J.: Indication of a universal persistence law governing atmospheric variability. Phys. Rev. Lett. **81**, 729–732 (1998)

Ladoy, P., Lovejoy, S., Schertzer, D.: Extreme variability of climatological data: scaling and intermittency. In: Non-linear Variability in Geophysics, pp. 241–250, Springer, Netherlands (1991)

Lavergnat, J., Gole, P.: A stochastic model of raindrop release: application to the simulation of point rain observations. J. Hidrol. **328**, 8–19 (2006)

Leonardo Fibonacci of Pisa, 1202: Fibonacci's Liber Abaci

Leopold, L.B., Miller, J.P.: Ephemeral streams—hydraulic factors and their relation to the drainage net. US Geol. Surv. Prof. Paper **282A** (1956)
Li, M., Zhao, W.: Golden ratio phenomenon of random data obeying von Karman spectrum. Math. Probl. Eng. **2013**, 1–6 (2013a). doi:10.1155/2013/130258
Li, M., Zhao, W.: Essay on Kolmogorov law of minus 5 over 3 viewed with Golden ratio. Adv. High Energy Phys. **2013**, 1–3 (2013b). doi:10.1155/2013/680678
Liebovitch, L.S., Scheurle, D.: Two lessons from fractals and chaos. Complexity **5**(4), 34–43 (2000)
Liu, S.H.: Formation and anomalous properties of fractals. IEEE Eng. Med. Biol. 28–39 (June, 1992)
Liu, Z., Xu, J., Shi, K.: Self-organized criticality of climate change. Theoret. Appl. Climatol. **115** (3–4), 685–691 (2014)
Liu, Z., Wang, L., Zhu, H.: A time-scaling property of air pollution indices: a case study of Shanghai, China. Atmos. Pollut. Res. **6**, 886–892 (2015)
Lovejoy, S.: Area-perimeter relation for rain and cloud areas. Science **216**, 185–187 (1982)
Lovejoy, S., Mandelbrot, B.B.: Fractal properties of rain, and a fractal model. Tellus **37A**, 209–232 (1985)
Lovejoy, S., Schertzer, D.: Scale invariance, symmetries, fractals, and stochastic simulations of atmospheric phenomena. Bull. Am. Meteorol. Soc. **67**, 21–32 (1986a)
Lovejoy, S., Schertzer, D.: Scale in variance in climatological temperatures and the local spectral plateau. Annates Geophysicae **86B**, 401–409 (1986b)
Lovejoy, S., Schertzer, D. (eds.): Nonlinear Variability in Geophysics: Scaling and Fractals. Kluwer, Dordrecht (1991)
Lovejoy, S., Schertzer, D.: Towards a new synthesis for atmospheric dynamics: space-time cascades. Atmos. Res. **96**, 1–52 (2010)
Mandelbrot, B.B.: The variation of certain speculative prices. J. Bus. **36**, 394–419 (1963)
Mandelbrot, B.B.: Very long-tailed probability distributions and the empirical distribution of city sizes. In: Massarik, F., Ratoosh, P. (eds.) Mathematical Explanations in Behavioral Science, pp. 322–332. Homewood Editions, New York (1965a)
Mandelbrot, B.B.: Time-Varying Channels, 1/*F* Noises and The Infrared Catastrophe, or: Why Does The Low Frequency Energy Sometimes Seem Infinite?. IEEE Communication Convention, Boulder CO (1965b)
Mandelbrot, B.B.: Self-similar error clusters in communications systems and the concept of conditional stationarity. IEEE Trans. Commun. Technol. **13**, 71–90. © Institute of Electrical and Electronics Engineers (1965c)
Mandelbrot, B.B., Van Ness, J.W.: Fractional brownian motions, fractional noises and applications. SIAM Rev. **10**, 422–437 (1968)
Mandelbrot, B.B., Wallis, J.R.: Noah, Joseph and operational hydrology. Water Resour. Res. **4**(5), 909–918 (1968)
Mandelbrot, B.B., Wallis, J.R.: Some long-run properties of geophysical records. Water Resour. Res. **5**(2), 321–340 (1969a)
Mandelbrot, B.B., Wallis, J.R.: Robustness of the rescaled range R/S in the measurement of noncyclic long-run statistical dependence. Water Resour. Res. **5**, 967–988 (1969b)
Mandelbrot, B.B.: Fractals: form, chance and dimension. W. H. Free-man and Co., San Francisco (1977)
Mandelbrot, B.B.: Fractals and Scaling in Finance. Springer, NewYork (1997)
Mandelbrot, B.B.: The Fractal Geometry of Nature. W. H. Free-man and Co., San Francisco (1983)
Mandelbrot, B.B., Hudson, R.L.: The (mis) behaviour of markets: a fractal view of risk, ruin and reward. Basic Books, New York (2004)
Markovic, D., Gros, C.: Power laws and self-organized criticality in theory and nature. Phys. Rep. **536**(2), 41–74. (2013) arxiv:1310.5527v3[nlin.AO]
Mauldin, R.D., Williams, S.C.: Random recursive construction. Trans. Am. Math. Sm. **295**, 325–346 (1986)

Miramontes, P., Li, W., Cocho, G.: Some critical support for power laws and their variations. (2012) arXiv:1204.3124v1[nlin.AO

Mitzenmacher, M.: A brief history of generative models for power law and lognormal distributions. Internet Math. **1**(2), 226–251 (2003)

Monetti, R.A., Havlin, S., Bunde, A.: Long-term persistence in the sea surface temperature fluctuations. Phys. A **320**, 581–589 (2003)

Montroll, E.W., Shlesinger, M.F.: On 1/*f* noise and other distributions with long tails. Proc. Natl. Acad. Sci. **79**, 3380–3383 (1982)

Montroll, E.W., Shlesinger, M.F.: Maximum entropy formalism, fractals, scaling phenomena, and 1/*f* noise: a tale of tails. J. Stat. Phys. **32**, 209–230 (1983)

Montroll, E.W., Shlesinger, M.F.: On the wonderful world of random walks. In: Lebowitz, J.L., Montroll, E.W. (eds.) Nonequilibrium Phenomena II, from Stochastic to Hydrodynamics, pp. 1–121. North Holland, Amsterdam (1984)

Nagel, K., Raschke, E.: Self-organized criticality in cloud formation? Phys. A **182**, 519–531 (1992)

Neelin, J.D., Peters, O., Hales, K.: The transition to strong convection. J. Atmos. Sci. **66**(8), 2367–2384 (2009)

Neelin, J.D., Peters, O., Lin, J.W.-B., Hales, K., Holloway, C.E.: Rethinking convective quasi-equilibrium: observational constraints for stochastic convective schemes in climate models. Phil. Trans. R. Soc. A **366**, 2581–2604 (2008)

Newman, M.E.J.: Power laws, Pareto distributions and Zipf's law. Contemp. Phys. **46**(5), 323–351 (2005)

Olsson, J., Niemczynowicz, J., Berndtsson, R.: Fractal analysis of high-resolution rainfall time series. J. Geophys. Res. **98**, 23265–23274 (1993)

Omori, F.: On the aftershocks of earthquakes. J. Coll. Sci. **7**, 111–200 (1894). (Imperial University of Tokyo)

Snell, O.: Die Abhängigkeit des Hirngewichts von dem Körpergewicht und den geistigen Fähigkeiten. Arch. Psychiatr. **23**(2), 436–446 (1892)

Pareto, V.: Cours d'Économie Politique Professé a l'Université de Lausanne, vol. I (1896)

Pareto, V.: Cours d'Économie Politique Professé a l'Université de Lausanne, vol. II (1897)

Pearson, K.: The Grammar of Science (2nd ed.). Adam and Charles Black, London (1900)

Peckham, S.D., Gupta, V.K.: A reformulation of Horton's law for large river networks in terms of statistical self- similarity. Water Resour. Res. **35**, 2763–2777 (1999)

Pelino, V., Matera, A., Colombo, T., Giorgi, F.: Validation of precipitation events in a regional climate model simulation using methods from complex systems theory. Theoret. Appl. Climatol. **84**(4), 213–218 (2006)

Peters, O., Hertlein, C., Christensen, K.: A complexity view of rainfall. Phys. Rev. Lett., **88**(1), 018701(1–4) (2002)

Peters, O., Neelin, J.D., Nesbitt, S.W.: Mesoscale convective systems and critical clusters. J. Atmos. Sci. **66**(9), 2913–2924 (2009)

Peters, O., Christensen, K.: Rain: relaxations in the sky. Phys. Rev. E **66**, 036120 (1–9) (2002), arXiv:cond-mat/0204109v1[cond-mat.soft]

Peters, O., Christensen, K.: Rain viewed as relaxation events. J. Hydrol. **328**(1–2), 46–55 (2006)

Peters, O., Neelin, J.D.: Critical phenomena in atmospheric precipitation. Nat. Phys. **2**(6), 393–396 (2006)

Peters, O., Deluca, A., Corral, A. Neelin, J.D., Holloway, C.E.: Universality of rain event size distributions. J. Stat. Mech. Theory Exp. **11**, P11030, (2010) arXiv:1010.4201v1[physics.ao-ph]

Pinto, C.M.A., Lopes, A.M., Machado, J.A.T.: A review of power laws in real life phenomena. Commun. Nonlinear Sci. Numer. Simulat. **17**, 3558–3578 (2012)

Popkov, V., Schadschneider, A., Schmidt, J., Schütz, G.M.: Fibonacci family of dynamical universality classes. PNAS **112**(41), 12645–12650 (2015)

Press, W.H.: Flicker noises in astronomy and elsewhere. Comments Mod. Phys. Part C-Comments Astrophys. **7**(4), 103–119 (1978)

Pruessner, G.: Studies in self-organised criticality. A thesis presented for the degree of Doctor of Philosophy of the University of London and the Diploma of Imperial College (2004)
Pruessner, G., Peters, O.: Self-organized criticality and absorbing states: lessons from the Ising model. Phys. Rev. E **73**(2), 025106(1–4) (2006)
Richardson, L.F.: Atmospheric diffusion shown on a distance-neighbor graph. Proc. R. Soc. London Ser. A **110**, 709–737 (1926)
Richardson, L.F.: 1960: The problem of contiguity: an appendix to statistics of deadly quarrels. In: Von Bertalanffy, L., Rapoport, A. (eds.) General Systems—Year Book of the Society for General Systems Research V, pp. 139–187. MI, Ann Arbor (1960)
Rybski, D., Bunde, A., Havlin, S., von Storch, H.: Longterm persistence in climate and the detection problem. Geophys. Res. Lett. **33**, L06718(1–4) (2006)
Samorodnitsky, G.: Long memory and self-similar processes. Annales de la faculté des sciences de Toulouse Mathématiques **15**(1), 107–123 (2006)
Samorodnitsky, G.: Long-range dependence. Found. Trends Stochast. Syst. **1**(3), 163–257 (2007)
Sarkar, A., Barat, P.: Analysis of rainfall records in India: self organized criticality and scaling. Fractals **14**(4), 289–293. (2006) arxiv.org/pdf/physics/0512197
Schick, K.L., Verveen, A.A.: 1/*f* noise with a low frequency white noise limit. Nature **251**, 599–601 (1974)
Schimper, C.F.: Geometrische Anordnung der um eine Axe periferischen Blattgebilde. Verhandl Schweiz Naturf Ges **21**, 113–117 (1836)
Selvam, A.M.: Deterministic chaos, fractals and quantumlike mechanics in atmospheric flows. Can. J. Phys. **68**, 831–841. (1990) http://xxx.lanl.gov/html/physics/0010046
Selvam, A.M., Pethkar, J.S., Kulkarni, M.K.: Signatures of a universal spectrum for atmospheric interannual variability in rainfall time series over the Indian Region. Int. J. Climatol. **12**, 137–152 (1992)
Selvam, A.M.: A universal spectrum for interannual variability of monsoon rainfall over India. Adv. Atmos. Sci. **10**(2), 221–226 (1993)
Selvam, A.M., Radhamani, M.: Signatures of a universal spectrum for nonlinear variability in daily columnar total ozone content. Adv. Atmos. Sci. **11**(3), 335–342 (1994)
Selvam, A.M., Radhamani, M.: Universal spectrum for short period (days) variability in atmospheric total ozone. Mausam **46**(3), 297–303 (1995)
Selvam, A.M., Joshi, R.R.: Universal spectrum for interannual variability in COADS global air and sea surface temperatures. Int J. Climatol. **15**, 613–623 (1995)
Selvam, A.M., Pethkar, J.S., Kulkarni, M.K.: Some unique characteristics of atmospheric interannual variability in rainfall time series over India and the United Kingdom. Adv. Atmos. Sci. **12**(3), 377–385 (1995)
Selvam, A.M., Pethkar, J.S., Kulkarni, M.K., Vijayakumar, R.: Signatures of a universal spectrum for atmospheric interannual variability in COADS surface pressure time series. Int J. Climatol. **16**, 393–404 (1996)
Selvam, A. M., Suvarna Fadnavis.: Signatures of a universal spectrum for atmospheric interannual variability in some disparate climatic regimes. Meteorol. Atmos. Phys. **66**, 87–112. (1998) http://xxx.lanl.gov/abs/chao-dyn/9805028
Selvam, A.M.: Chaotic Climate Dynamics. Luniver Press, UK (2007)
Selvam, A.M.: Fractal fluctuations and statistical normal distribution. Fractals **17**(3), 333–249. (2009) http://arxiv.org/pdf/0805.3426
Selvam, A.M.: Signatures of universal characteristics of fractal fluctuations in global mean monthly temperature anomalies. J. Syst. Sci. Complex. **24**(1), 14–38 (2011a)
Selvam, A.M.: A general systems theory for atmospheric flows and atmospheric aerosol size distribution. Chaotic Modeling and Simulation (CHAOS 2011 Conference Proceedings) 461–468. (2011b) http://arxiv.org/ftp/arxiv/papers/0908/0908.2321.pdf
Selvam, A.M.: Universal spectrum for atmospheric suspended particulates: comparison with observations. Chaos Complex. Lett. **6**(3), 1–43. (2012a) http://arxiv.org/abs/1005.1336

Selvam, A.M.: Universal spectrum for atmospheric aerosol size distribution: comparison with pcasp-b observations of vocals 2008. Nonlinear Dyn. Syst. Theor. **12**(4), 397–434. (2012b) http://arxiv.org/abs/1105.0172

Selvam, A.M.: Scale-free universal spectrum for atmospheric aerosol size distribution for Davos, Mauna Loa and Izana. Int. J. Bifurcation Chaos **23**, 1350028 (1–13). (2013) http://arxiv.org/pdf/1111.3132

Selvam, A.M.: Universal inverse power-law distribution for temperature and rainfall in the UK region. Dyn. Atmos. Oceans **66**, 138–150 (2014)

Selvam, A.M.: Rain Formation in Warm Clouds: General Systems Theory. SpringerBriefs in Meteorology, Springer (2015)

Smoluchowski, M.: Essai d'une théorie cinétique du mouvement Brownien et des milieux troubles (Outline of the kinetic theory of Brownian motion of suspensions). Bulletin International de l'Académie des Sciences de Cracovie, 577–602 (1906)

Sornette, D.: Critical Phenomena in Natural Sciences. Springer Series in Synergetics (2006)

Sornette, D.: Probability distributions in complex systems. In: Meyers, R.A. (ed.) Computational Complexity, pp. 2286–2300. Springer, New York (2012). http://arxiv.org/pdf/0707.2194v1.pdf

Sornette, D., Ouillon, G.: Dragon-kings: mechanisms, statistical methods and empirical evidence. Eur. Phys. J. Spec. Top. **205**, 1–26 (2012)

Stumpf, M.P.H., Porter, M.A.: Critical truths about power laws. Science **335**(6069), 665–666 (2012)

Thompson, D.W.: On Growth and Form. Cambridge Univ. Press, Cambridge (1942). (The first edition appeared in 1917)

Turcotte, D.L.: Self-organized criticality. Rep. Prog. Phys. **62**, 1377–1429 (1999)

Van der Hoven, I.: Power spectrum of the horizontal wind speed in the frequency range from 0.0007 to 900 cycles per hour. J. Meteor. **14**, 160–164 (1957)

Van Der Ziel, A.: On the noise spectra of semi-conductor noise and of flicker effect. Physica **16**(4), 359–372 (1950)

Vattay, G., Harnos, A.: Scaling behavior in daily air humidity fluctuations. Phys. Rev. Lett. **73**(5), 768–771 (1994)

Vinnichenko, N.K.: The kinetic energy spectrum in the free atmosphere, 1 second to 5 years. Tellus **22**(2), 158–166 (1970)

von Karman, T.: Progress in the statistical theory of turbulence. Proc. Nat. Acad. Sci. **34**(11), 530–539 (1948)

von Kármán, Th.: Mechanische Ähnlichkeit und Turbulenz. Nachrichten von der Gesellschaft der Wissenschaften zu Göttingen, Fachgruppe 1 (Mathematik) **5**, 58–76 (also as: Mechanical Similitude and Turbulence, Tech. Mem. NACA, no. 611, 1931) (1930)

Wang, G., Jiang, T., Blender, R., Fraedrich, K.: Yangtze 1/*f* discharge variability and the interacting river-lake system. J. Hydrol. **351**, 230–237 (2008)

Wang, Z., Huang, C.: Self-organized criticality of rainfall in central china. Adv. Meteorol. **203682** (1–8) (2012)

Watkins, N.W., Pruessner, G., Chapman, S.C., Crosby, N.B.: 25 Years of SOC: concepts and controversies. Space Sci. Rev. **198**(1), 3–44 (2016)

Weissman, M.B.: 1/*f* noise and other slow, nonexponential kinetics in condensed matter. Rev. Mod. Phys. **60**, 537–571 (1988)

West, B.J.: Colloquium: fractional calculus view of complexity: a tutorial. Rev. Mod. Phys. **86**, 1169–1184 (2014)

West, G.B., Brown, J.H.: The origin of allometric scaling laws in biology from genomes to ecosystems: towards a quantitative unifying theory of biological structure and organization. J. Exp. Biol. **208**, 1575–1592, Published by The Company of Biologists 2005, (2005). doi:10.1242/jeb.01589

West, G.B., Brown, J.H., Enquist, B.J.: A general model for the origin of allometric scaling laws in biology. Science **276**, 122–126 (1997)

West, G.B., Brown, J.H., Enquist, B.J.: The fourth dimension of life: Fractal geometry and allometric scaling of organisms. Science **284**, 1677–1679 (1999)

Willinger, W., Alderson, D., Doyle, J. C., Li, L.: More “normal” than normal: Scaling distributions and complex systems. In: Ingalls, R.G., Rossetti, M.D., Smith, J.S., Peters, B.A. (eds.) Proceedings of the 2004 Winter Simulation Conference, pp. 130–141 (2004)
Wilson, E.B., Hilferty, M.M.: Note on C.S. Peirce’s experimental discussion of the law of errors. Proc. Natl. Acad. Sci. U.S.A. **15**(2), 120–125 (1929)
Wilson, K.G.: Problems in physics with many scales of length. Sci. Am. **241**(2), 158–179 (1979)
Wilson, K.G.: The renormalization group and critical phenomena. Nobel lecture, 8 December 1982 by Kenneth G. Wilson. (1982) http://www.nobelprize.org/nobel_prizes/physics/laureates/1982/wilson-lecture.pdf
Wood, R., Field, P.: The distribution of cloud horizontal sizes. J. Clim. **24**, 4800–4816 (2011)
Yano, J.-I., Guichard, F., Lafore, J.-P., Redelsperger, J.-L., Bechtold, P.: Renormalization approach for subgrid-scale representations. J. Atmos. Sci. **60**, 2029–2038 (2003)
Yano, J.-I., Nishi, N.: The hierarchy and self-affinity of the time-variability within the tropical atmosphere inferred from the NOAA OLR data. J. Meteorol. Soc. Jpn **67**, 771–789 (1989)
Yano, J.-I., Blender, R., Zhang, C., Fraedrich, K.: $1/f$—noise and pulse-like events in the tropical atmospheric surface variabilities. Q. J. Roy. Meteor. Soc. **130**, 1697–1721 (2004)
Yano, J.-I., Fraedrich, K., Blender, R.: Tropical convective variability as $1/f$ noise. J. Climate **14**, 3608–3616 (2001)
Zeng, L., Wang, G.: Modeling golden section in plants. Prog. Nat. Sci. **19**, 255–260 (2009)
Zipf, G.K.: Human Behavior and the Principle of Least Effort. Addison-Wesley, Cambridge, MA (1949)

Chapter 3
Self-organized Criticality: A Signature of Quantum-like Chaos in Atmospheric Flows

Abstract Atmospheric flows exhibit long-range spatiotemporal correlations manifested as the fractal geometry to the global cloud cover pattern concomitant with inverse power-law form for power spectra of temporal fluctuations on all space-time scales ranging from turbulence (centimetres-seconds) to climate (kilometres-years). Long-range spatiotemporal correlations are ubiquitous to dynamical systems in nature and are identified as signatures of *self-organized criticality*. Standard models in meteorological theory cannot explain satisfactorily the observed self-organized criticality in atmospheric flows. Mathematical models for simulation and prediction of atmospheric flows are nonlinear and do not possess analytical solutions. Finite precision computer realizations of nonlinear models give unrealistic solutions because of *deterministic chaos*, a direct consequence of round-off error growth in iterative numerical computations. Recent studies show that round-off error doubles on an average for each iteration of iterative computations. Round-off error propagates to the mainstream computation and gives unrealistic solutions in numerical weather prediction (NWP) and climate models, which incorporate thousands of iterative computations in long-term numerical integration schemes. A general systems theory model for atmospheric flows developed by the author predicts the observed *self-organized criticality* as intrinsic to quantumlike chaos in flow dynamics. The model provides universal quantification for *self-organized criticality* in terms of the golden ratio τ (≈1.618). Model predictions are in agreement with a majority of observed spectra of time series of several standard meteorological and climatological data sets representative of disparate climatic regimes. Universal spectrum for natural climate variability rules out linear trends. Man-made greenhouse gas related atmospheric warming would result in intensification of natural climate variability, seen immediately in high-frequency fluctuations such as QBO and ENSO and even shorter timescales. Model concepts and results of analyses are discussed with reference to possible prediction of climate change. Model concepts, if correct, rule out unambiguously, linear trends in climate. Climate change will only be manifested as increase or decrease in the natural variability. However, more stringent tests of model concepts and predictions are required before applications to such an important issue as *climate change*. The cell dynamical system model for atmospheric flows is a general systems theory

A.M. Selvam, *Self-organized Criticality and Predictability in Atmospheric Flows*,
Springer Atmospheric Sciences, DOI 10.1007/978-3-319-54546-2_3

applicable in general to all dynamical systems in other fields of science, such as, *number theory*, *biology*, *physics* and *botany*.

Keywords General systems theory · Nonlinear dynamics and chaos · Fractals · Long-range space-time correlations · Self-organized criticality · Quantumlike chaos

3.1 Introduction

Turbulence, namely, irregular fluctuations in space and time characterize fluid flows in general and atmospheric flows in particular. The irregular, i.e., nonlinear space-time fluctuations on all scales contribute to the unpredictable nature of both short-term weather and long-term climate. Quantification of atmospheric flow patterns as recorded by meteorological parameters such as temperature, wind speed, pressure, etc., will help exact prediction of weather and climate and also provide a model for turbulent fluid flows in general. Meteorologists have documented in detail the nonlinear variability of atmospheric flows, in particular, the interannual variability, i.e., the year-to-year fluctuations in weather patterns. The interannual variability of atmospheric flows is nonlinear and exhibits fluctuations on all scales ranging up to the length of data period (time) investigated. The broadband spectrum of atmospheric interannual variability has embedded dominant quasiperiodicities such as the quasibiennial oscillation (*QBO*) and the *ENSO* (El Nino/Southern Oscillation) cycle of 3 to 7 years (Philander 1990) which are identified as major contributors to local climate variability, in particular, the monsoons which influence agriculture dependent world economies. *ENSO* is an irregular (3–7 years), self—sustaining cycle of alternating warm and cool water episodes in the Pacific Ocean. Also called *El Nino—La Nina*, *La Nina* refers to the cool part of the weather cycle while *El Nino* is associated with a reversal of global climatic regimes resulting in anomalous floods and droughts throughout the globe. It is of importance to quantify the total pattern of fluctuations for predictability studies. Observations show that atmospheric flows exhibit fluctuations on all scales (space-time) ranging from turbulence (mm-sec) to planetary scale (thousand of kilometres-year). The power spectra of temporal fluctuations are broadband and exhibit inverse power-law form $1/f^{\alpha}$ where f is frequency and α, the exponent, is different for different scale ranges. Inverse power-law form for power spectra implies scaling (self-similarity) for the scale range over which α is constant. Atmospheric flows therefore exhibit multiple scaling or *multifractal* structure. The *fractal* and *multifractal* nature of fluid turbulence in general and also in atmospheric flows has been discussed in detail by Sreenivasan (1991). The word *fractal* was first coined by Mandelbrot (1977, 1983) to describe the self-similar fluctuations that are generic to dynamical evolution of systems in nature. *Fractals* signify non-*Euclidean* or fractional *Euclidean* geometrical structure. Traditional statistical theory does not provide for a satisfactory description and quantification of such nonlinear variability with multiple scaling.

The apparently chaotic nonlinear variability (intermittency) of atmospheric flows therefore exhibit implicit order in the form of multiple scaling or *multifractal* structure of temporal fluctuations implying long-range temporal correlations, i.e., the amplitudes of long-term and short-term fluctuations are related by a multiplication factor proportional to the scale ratio and therefore independent of exact details of dynamical evolution of fluctuations (Schroeder 1991; Stanley 2000). Recent studies (since 1988) in all branches of science reveal that self-similar *multifractal* spatial pattern formation by self-similar fluctuations on all space-time scales is generic to dynamical systems in nature and is identified as signature of *self-organized criticality* (Bak et al. 1987, 1988; Bak and Chen 1989, 1991; Buchanan 1997). Such *multifractal* temporal fluctuations in atmospheric flows are associated with self-similar *multifractal* spatial patterns for cloud and rain areas documented and discussed in great detail by Lovejoy and his group (Lovejoy and Schertzer 1986a, b; Schertzer and Lovejoy 1991, 1994; Tessier et al. 1993). Standard meteorological theory cannot explain satisfactorily the observed *multifractal* structure of atmospheric flows. Self-similar spatial pattern implies long-range spatial correlations. Atmospheric flows therefore exhibit long-range spatiotemporal correlations, namely, *self-organized criticality*, signifying order underlying apparent chaos. Prediction may therefore be possible. Statistical prediction models are based on observed correlations, which, however, change with time, thereby introducing uncertainties in the predictions. Traditionally, prediction of atmospheric flow patterns has been attempted using mathematical models of turbulent fluid flows based on *Newtonian* continuum dynamics. Such models are nonlinear and finite precision computer realizations give chaotic solutions because of sensitive dependence on initial conditions, now identified as *deterministic chaos*, an area of intensive research in all branches of science since 1980 (Gleick 1987). Sensitive dependence on initial conditions in computed solutions implies long-range spatiotemporal correlations, namely *self-organized criticality*, similar to that observed in real world dynamical systems. *Deterministic chaos* in computed solutions precludes long-term prediction. The fidelity of computed solutions is questionable in the absence of analytical (true) solutions (Stewart 1992, 2011). Systematic round-off error that can be encountered in numerical computations has been discussed (Vivaldi 1994; Konar 2006; Wang et al. 2006). *Deterministic chaos* is a direct consequence of round-off error growth in finite precision computer solutions of error sensitive dynamical systems such as $X_{n+1} = F(X_n)$, where X_{n+1}, the $(n + 1)$th value of the variable X at the $(n + 1)$th instant is a function F of X_n. Selvam (1993) has shown that round-off error approximately doubles on an average for each iteration in iterative computations and give unrealistic solutions in numerical weather prediction (*NWP*) and climate models which incorporate thousands of iterations in long-term numerical integration schemes. Computed model solutions are therefore mere mathematical artifacts of the universal process of round-off error growth in iterative computations. Selvam (1993) has shown that the computed domain is the successive cumulative integration of round-off error domains analogous to the formation of large eddy domains as envelopes enclosing turbulent eddy fluctuation domains such as in atmospheric flows (Selvam 1990;

Selvam et al. 1992, 1996; Selvam and Joshi 1995). Computed solutions, therefore qualitatively resemble real world dynamical systems such as atmospheric flows with manifestation of *self-organized criticality*. *Self-organized criticality*, i.e. long-range spatiotemporal correlations, originates with the primary perturbation domains corresponding respectively to round-off error and dominant turbulent eddy fluctuations in model and real world dynamical systems. Computed solutions, therefore, are not true solutions. The vast body of literature investigating chaotic trajectories in recent years (since 1980) document, only the round-off error structure in finite precision computations. The physical mechanism underlying *self-organized criticality* in model and real world dynamical systems is not yet identified. A review of literature relating to studies on *Nonlinear Dynamics and Chaos* and applications for prediction for weather and climate is given in Zeng et al. (1993) and Selvam and Fadnavis (1998). A review of research work on the applications of self-similarity and self-organized criticality in atmospheric sciences is given in Dessai and Walters (2000). Watkins et al. (2016) have given current status of applications of the concept of self-organized criticality in all branches of science.

The author has developed a general systems theory applicable to all dynamical systems including atmospheric flows (Selvam 1990, 2007, 2015; Selvam and Fadnavis 1998) which predicts the observed *self-organized criticality* as intrinsic to quantumlike mechanics governing flow dynamics. *Self-organized criticality* is therefore a signature of *quantumlike chaos* in dynamical systems.

The model concepts are based on the observational characteristic that real world dynamical systems such as fluid flows, heart beat patterns, etc., exhibit self-similar fractal fluctuations. The power (variance) spectra of fractal fluctuations exhibit inverse power-law form $f^{-\alpha}$ where f is the frequency and α the exponent and signify a scale invariant eddy continuum for the range of frequencies for which α is a constant, i.e. the amplitudes of component eddies are related to each other by the scale factor α alone independent of other dynamical processes governing the growth processes of the eddies (waves). The cooperative existence of eddies in the scale invariant eddy continuum exhibit long-range space-time correlations or persistence between the fluctuation amplitudes of larger and smaller scale eddies.

Further, the fractal space-time structure to the string-like atmospheric flow trajectory is resolved into a continuum of eddies. The eddy circulations obey *Kepler*'s third law of planetary motion and therefore eddy inertial masses obey *Newton's inverse square law of gravitation* on all scales from microscopic to macroscale. El Naschie (Iovane et al. 2008) has discussed, in a series of publications, the fractal structure to space-time and also states that fractalisation of microspace is the origin of gravity.

The model provides universal quantification for *self-organized criticality* in terms of the universal inverse power-law form which is a function of the golden ratio ($\approx$1.618) and is close to the statistical normal distribution for the power spectra of space-time fluctuations. The model predictions are in agreement (see all references under Selvam) with continuous periodogram spectral analysis of meteorological data sets. The model concepts are independent of the exact details such as the chemical, physical, physiological and other properties of the dynamical system and therefore provide a general systems theory applicable to all real world and computed dynamical systems in nature (von Bertalanffy 1972; Selvam 1998, 1999, 2001a, b, 2002a, b; Selvam et al. 2000).

3.2 Model Concepts and Predictions

a. The general systems theory model visualizes the generation of scale invariant eddy continuum in turbulent fluid flows by spontaneous organized growth of large eddies from space-time integration of enclosed small-scale eddies in successive stages. The eddy continuum exists as a unified whole communicating network with ordered energy flow between the larger and smaller scale eddies.
b. Space-time integration of enclosed small scale (turbulent) eddies of radius r and circulation speed w_* gives rise to large eddy of radius R associated with circulation speed W and the relationship between the radii and circulation speeds of the large and small eddies is given as (Selvam 1990)

$$W^2 = \frac{2}{\pi}\frac{r}{R}w_*^2 \tag{3.1}$$

The variance ratio W^2/w_*^2 represents the probability distribution r/R of fractal fluctuations from Eq. 3.1. Additive amplitudes of eddies when squared represent the probability distribution of eddy variance is similar to quantum mechanical laws governing the subatomic dynamics of quantum systems such as the electron or photon (Maddox 1988). Incidentally, one of the strangest things about physics is that we seem to need two different kinds of mechanics, quantum mechanics for microscopic dynamics of quantum systems and classical mechanics for macro-scale phenomena (Maddox 1993; Rae 1988). The above visualization of the unified network of atmospheric flows as a quantum system is consistent with Grossing's (1989) concept of quantum systems *as order out of chaos* phenomena. Order and chaos have been reported in strong fields in quantum systems (Brown 1996).
c. The square of the eddy amplitude W^2 represents the kinetic energy E given as (from Eq. 3.1) $E = H\nu$ where ν (proportional to $1/R$) is the frequency of the large eddy and H is a constant equal to $\frac{2}{\pi}rw_*^2$ for growth of large eddies sustained by constant energy input proportional to w_*^2 from fixed primary small scale eddy fluctuations. Energy content of eddies is therefore similar to quantum systems which can possess only discrete quanta or packets of energy content $h\nu$ where h is a universal constant of nature (Planck's constant) and ν is the frequency in cycles per second of the electromagnetic radiation. The relative phase angle between large and turbulent eddies is equal to r/R and is directly proportional to W^2 (Eq. 3.1). The phase angle therefore represents variance and also there is

progressive increase in phase with increase in wavelength. The above relationship between phase angle, variance and frequency has been identified as *Berry's Phase* (Berry 1988) in the subatomic dynamics of quantum systems. Berry's phase has been identified in atmospheric flows (Selvam and Fadnavis 1998).

d. Writing Eq. (3.1) in terms of the periodicities T and t of large and small eddies respectively, where

$$T = \frac{2\pi R}{W} \text{ and } t = \frac{2\pi r}{w_*}$$

we obtain
$$\frac{R^3}{T^2} = \frac{2}{\pi}\frac{r^3}{t^2} = \text{constant} \quad (3.2)$$

Equation (3.2) is analogous to *Kepler*'s third law of planetary motion, namely, the square of the planet's year (period) to the cube of the planet's mean distance from the *Sun* is the same for all planets (Weinberg 1993). Newton developed the idea of an inverse square law for gravitation in order to explain Kepler's laws, in particular, the third law. Kepler's laws were formulated on the basis of observational data and therefore are of empirical nature. A basic physical theory for the inverse square law of gravitation applicable to all objects, from macroscale astronomical objects to microscopic scale quantum systems is still lacking. The model concepts are analogous to a string theory (Kaku 1997) where, superposition of different modes of vibration in stringlike energy flow patterns result in material phenomena with intrinsic quantumlike mechanical laws which incorporate inverse square law for inertial forces, the equivalent of gravitational forces, on all scales of eddy fluctuations from macro- to microscopic scales. The cumulative sum of centripetal forces in a hierarchy of vortex circulations may result in the observed inverse square law form for gravitational attraction between inertial masses (of the eddies). Uzer et al. (1991) have discussed new developments within the last two decades which have spurred a remarkable revival of interest in the application of classical mechanical laws to quantum systems. The atom was originally visualized as a miniature solar system based on the assumption that the laws of classical mechanics apply equally to electrons and planets. However within a short interval of time the new quantum mechanics of *Schrodinger* and *Heisenberg* became established (from the late 1920s) and the analogy between the structure of the atom and that of the solar system seemed invalid and classical mechanics became the domain of the astronomers. There is now a revival of interest in classical and semiclassical methods which are found to be unrivalled in providing an intuitive and computationally tractable approach to the study of atomic, molecular and nuclear dynamics.

The apparent paradox of wave–particle duality in microscopic scale quantum systems (Rae 1988) is however physically consistent in the context of macroscale atmospheric flows since the bidirectional energy flow structure of a complete atmospheric eddy results in the formation of clouds in updraft regions and dissipation of clouds in downdraft regions. The commonplace occurrence of clouds in a row is a manifestation of *wave–particle duality* in the macroscale quantum system of atmospheric flows (Figs. 3.1 and 3.2).

The above-described analogy of quantumlike mechanics for atmospheric flows is similar to the concept of a subquantum level of fluctuations whose space-time organization gives rise to the observed manifestation of subatomic phenomena, i.e. quantum systems as order out of chaos phenomena.

Puthoff (1987) has shown that the observed stability of ground-state electronic orbits in atoms is a result of energy exchange with the sea of electromagnetic energy available in the vacuum zero-point fluctuations. Historically, quantum mechanics had imposed arbitrary stability criterion for the ground state of electron orbits. Stable ground state is not possible in classical physics since attractive forces between the negative electron and positive nucleus will result in spiralling of orbital electrons into the nucleus accompanied by loss of energy due to emission of radiation by the accelerating electron, since all accelerating charges radiate energy. Puthoff (1989) has also put forth the concept of "gravity as a zero-point fluctuation force". The vacuum zero-point fluctuation (electromagnetic) energy is manifested in the *Casimir effect* (Spruch 1996), namely a force between two closely spaced metal plates. Casimir effect is interpreted as due to imbalances in the zero-point energy caused by the presence of the plates and is analogous to the turbulent scale fluctuations whose spatial integration results in coherent large eddy structures. Recent studies show that background noise enhances weak signals in electronic circuits (Moss and Wiesenfeld 1995). El Naschie has proposed in a series of papers (Iovane et al. 2008) that *Cantorian-fractal* conception of space-time may effect reconciliation between quantum mechanics and gravity.

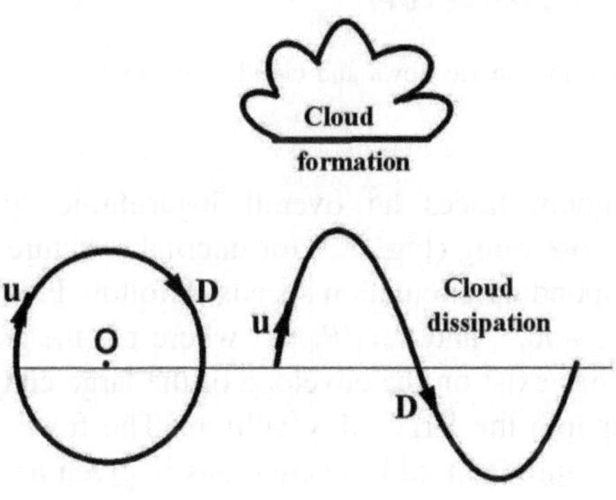

Fig. 3.1 Instantaneous non-local connection in atmospheric eddy circulations

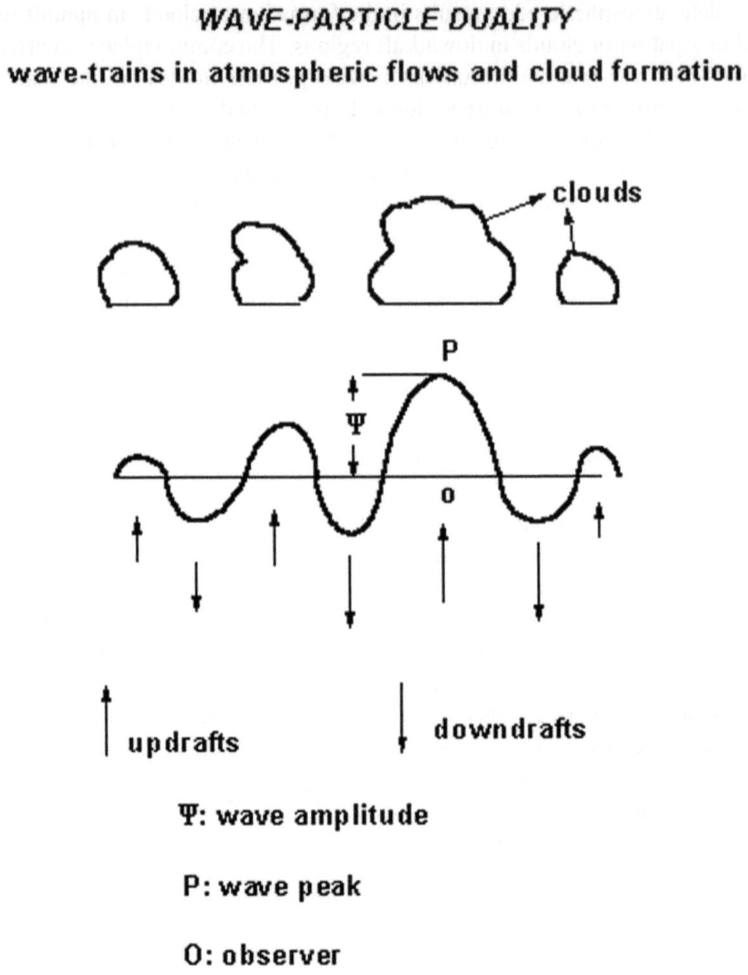

Fig. 3.2 Wave trains in atmospheric flows and cloud formation

e. The eddy continuum traces an overall logarithmic spiral trajectory with quasiperiodic Penrose tiling (Fig. 3.3) for internal structure such that successive radii R and corresponding circulation speeds W follow Fibonacci number series, namely $R_{n+1} = R_n + R_{n-1}$ and $R_{n+1}/R_n = \tau$ where τ is the golden ratio ($\approx$1.618).
f. The turbulent eddies exist on the envelope of the large eddy and mix overlying environmental air into the large eddy volume. The fractional volume dilution k of large eddy by turbulent eddy fluctuations is given as

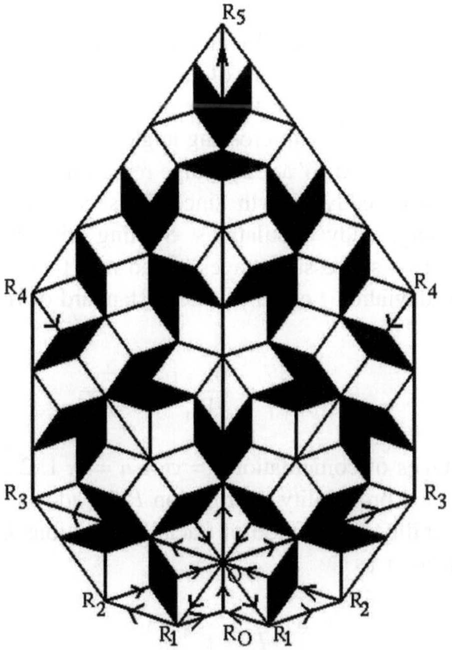

Fig. 3.3 The internal structure of large eddy circulations

$$k = \frac{w_* r}{WR} \tag{3.3}$$

The steady state dominant eddies R_OOR_1, R_1R_2O, R_3R_2O, R_3R_4O, of the quasiperiodic Penrose tiling pattern do not get erased by turbulent eddy fluctuations since the fractional volume dilution k by turbulent eddy mixing equal is to $1/\tau^2 \approx 0.382$, i.e. less than half.

g. The logarithmic spiral eddy flow structure is given in terms of the eddy length scale z equal to R/r and the corresponding circulation speeds W and w_* as

$$W = \frac{w_*}{k} \ln z \tag{3.4}$$

von Karman (1930) showed that turbulent fluid flow dynamics near a wall follow the 'law of the wall', namely the average velocity of a turbulent flow at a certain point is proportional to the logarithm of the distance from that point to the "wall", or the boundary of the fluid region. The logarithmic law of the wall derived by von Karman is the same as Eq. (3.4) with k a non-dimensional constant named von Karman's constant equal to about 0.4 obtained from experiments. Recently, Li and Zhao (2013a, b) have shown that the power spectra for turbulent fluid flows derived by Kolmogorov (1941) and von Karman (1948) exhibit self-similar, i.e. fractal character of turbulent eddy fluctuations and can be expressed in terms of the golden ratio. Therefore the trajectory of turbulent fluid flows near a boundary or wall follows a logarithmic spiral with the golden ratio as the winding number similar to the spiralling arrangement of successive leaves in plant growth (Jean 1994). Selvam (1990) has shown that the atmospheric flow structure follows a logarithmic spiral trajectory with crossing angle equal to the golden mean.

h. The r.m.s circulation speed W at each stage represents the standard deviation σ for the next stage of eddy growth since W is computed as the average of enclosed earlier stage eddy circulations. Starting with the same primary perturbation w_*, for two successive stages n and $n + 1$ of eddy growth the normalized standard deviation t equal to (mean/standard deviation) is given as

$$t = \frac{\sigma_{n+2}}{\sigma_{n+1}} = \frac{W_{n+1}}{W_n} = \frac{n+1}{n} \tag{3.5}$$

For successive stages of computation $t = \sigma = n = 0, 1, 2, \ldots n$.

i. The model predicted probability distribution P for eddy variance spectrum and also for the probability distribution of fractal fluctuations is given as a function of the golden mean τ as

$$P = \tau^{-4t} \tag{3.6}$$

Therefore fractal fluctuations signify quantum-like chaos since the property that the additive amplitude of eddies when squared represent the probability densities is exhibited by the subatomic dynamics of quantum systems such as the electron or photon (Maddox 1988) (a brief review of current status quantum mechanics is given in Sect. 3.5).

The normalized standard deviation t equal to $(x–xbar)/sdev$ where $xbar$ and $sdev$ are the respective mean and standard deviation of the fractal fluctuations x is used to compute the probability P values for the region of dominant large eddy growth, namely $-1 \geq t \geq 1$. The probability P for the region of primary eddy growth, namely $-1 \leq t \leq 1$ is computed using the equation $P = \tau^{-4k}$ where $k = wr/WR$ is the fractional volume dilution of the growing primary eddy by turbulent eddy fluctuations.

3.3 Deterministic Chaos and Statistical Normal Distribution

The statistical distribution characteristics of fluctuations in natural phenomena follow normal distribution associated conventionally with random chance. The normal distribution is characterized by (1) the moment coefficient of skewness equal to zero, signifying symmetry and (2) the moment coefficient of kurtosis equal to 3 representing intermittency of turbulence on relative time scale. In the following it is shown that the universal period doubling route to chaos growth phenomena in nature generates distribution characteristics very close to the observed statistical normal distribution parameters. The period doubling route to growth is initiated and sustained by the turbulent (fine scale) eddy acceleration w_* which then propagates by the inherent property of the inertia of the medium. In the context of atmospheric turbulence, the statistical parameters, mean, variance, skewness and kurtosis represent respectively the net vertical velocity, intensity of turbulence, vertical momentum flux and intermittency of turbulence and are given respectively by w_*, w_*^2, w_*^3, w_*^4. By analogy, the perturbation speed w_* (motion) per second of the medium sustained by its inertia represents the mass; w_*^2, the acceleration (or force); w_*^3, the momentum (or potential energy) and w_*^4, the spin angular momentum, since an eddy motion is inherently symmetric with bidirectional energy flow, the skewness factor w_*^3 is equal to zero for one complete eddy circulation thereby satisfying the law of conservation of momentum. The moment coefficient of kurtosis which represents the intermittency of turbulence is shown in the following to be slightly less than that for normal distribution value 3 for the period doubling route to chaos eddy growth phenomena. Selvam (2013) has shown that the probability distribution of fractal fluctuations is very close to the statistical normal distribution for values of normalized standard deviation $t \leq 2$ and gives appreciably larger probability values for larger standard deviations

The numerical value of the fourth moment about the mean which represents the statistical intermittency of turbulence as well as the dynamical spin angular momentum of the eddy for the eddy growth process at Eq. 3.1 is computed in the following. From Eq. (3.3), namely $k = \frac{wr}{WR}$

$$\mathrm{d}W = \frac{w_*}{k}\mathrm{d}\ln z$$

The fourth moment about the mean is given as

$$\left(\frac{\mathrm{d}W}{w_*}\right)^4 = \left(\frac{\mathrm{d}\ln z}{k}\right)^4 = \left(\frac{\mathrm{d}z}{zk}\right)^4 = \left(\frac{r}{Rk}\right)^4 \tag{3.7}$$

The fourth moment about the mean $\left(\frac{\mathrm{d}W}{w_*}\right)^4$ represents the relative statistical coefficient of kurtosis and also the relative spin angular momentum for the eddy dynamical growth process. Organized eddy growth occurs for scale ratio equal to 10 and identifies the large eddy on whose envelope period doubling growth process occurs. Therefore, for a dominant eddy

$$\frac{r}{R} = \frac{1}{10}$$

From Eq. 3.1

$$\frac{w_*}{W} = \sqrt{\frac{\pi}{2}\frac{R}{r}}$$

From Eq. 3.3

$$k = \frac{w_* r}{WR} = \sqrt{\frac{\pi}{2}\frac{R}{r}\frac{r}{R}} = \sqrt{\frac{\pi}{2}\frac{r}{R}} = \sqrt{\frac{\pi}{20}} \approx 0.4$$

The computed value of k, the steady state fractional volume dilution of large eddy by turbulent eddy mixing and which is also the same as von Karmen's constant in Eq. 3.4 (Sect. 3.2 list item *f*) is in agreement with experimentally measured value equal to about 0.4.

$(dz/z) = 1/2$ for one length growth by period doubling process since $z = dz + dz$ where z is the length scale ratio equal to R/r. Therefore moment coefficient of kurtosis (Eq. 3.7) is equal to

$$\left(\frac{dW}{w_*}\right)^4 = \left(\frac{r}{Rk}\right)^4 = \frac{1}{2^4}\left(\frac{20 \times 7}{22}\right)^2 = 2.53 \tag{3.8}$$

Moment coefficient of kurtosis (Eq. 3.8) equal to 2.53 is less than that for the normal distribution value 3.0. Selvam (2013) has shown that probability distribution of fractal fluctuations is close to normal distribution for small to moderate fluctuations (up to 2 standard deviations) but exhibits a fat long tail for larger fluctuations. Period doubling growth phenomena results in a 2.53 times increase in the spin angular momentum of the large eddy for each period doubling sequence. Period doubling at constant pump frequency involves eddy length step growth dR on either side of the primary turbulent eddy length dR.

3.4 Atmospheric Eddy Growth Process as Universal Period Doubling Route to Chaos

A general systems theory model for turbulent fluid flows such as atmospheric flows developed by Selvam (1990, 2007, 2009, 2014, 2015) predicts that the observed self-organized criticality in real world dynamical systems is a manifestation of the universal period doubling route to chaos quantified by the Feigenbaum's constants in computer realizations of nonlinear dynamical systems investigated extensively in the newly emerging field of 'nonlinear dynamics and chaos' in the 1970s and early 1980s (Ecke 2015).

The universal period doubling route to chaos has been studied extensively by mathematicians, the basic example with the potential to display the main features of the erratic behaviour is the Julia model (Delbourgo 1986) given below.

$$X_{n+1} = F(X_n) = LX_n(1 - X_n)$$

The above nonlinear model represents the population values of the parameter X at different time periods n and L parameterises the rate of growth of X for small X.

The Eq. (3.1), namely $W^2 = \frac{2}{\pi}\frac{r}{R}w_*^2$ representing large eddy growth (radius R and circulation speed W) as integrated space-time mean of turbulent eddy fluctuation (radius r and circulation speed w_*) is analogous to the Julia model since large eddy growth is dependent on the energy input from the turbulence scale with ordered two-way energy feedback between the larger and smaller scales (Selvam 1993).

Feigenbaum's (1980) research showed that the successive spacing ratios of X and L for adjoining period doublings are given respectively by the two universal constants $a = -2.5029$ and $d = 4.6692$. The universal constants a and d assume different numerical values for period tripling, quadrupling etc. and the appropriate values are computed by Delbourgo (1986) and shown to follow the relation $3d = 2a^2$ over a wide domain.

The physical concept of large eddy growth by the period doubling process enables to derive the universal constants a and d and their mutual relationship as functions inherent to the scale invariant eddy energy structure as follows (Selvam 2009).

3.4.1 *Feigenbaum's Constant* a

The Feigenbaum's constant a is given by the successive spacing ratios W for adjoining period doublings. W and R are respective successive spacing intervals, because by definition W and R are computed as incremental growth steps dW and dR for each stage of computation. Earlier (Selvam 1990) it was shown that successive values of eddy radii and corresponding circulation speeds follow Fibonacci mathematical number series such that the ratio of successive values of W and R is equal to the golden ratio τ ($\approx$1.618).

The Feigenbaum's constant a is obtained as the successive spacing ratios of W, i.e.

$$a = \frac{w_*}{W_1} = \frac{W_1}{W_2} = \frac{W_2}{W_3} \tag{3.9}$$

For the same primary perturbation w_* of radius r_* the large eddy domains W_1R_1 and W_2R_2 are related to each other as (Eq. 3.1)

$$W_1^2 = \frac{2}{\pi}\frac{r}{R_1}w_*^2$$
$$W_2^2 = \frac{2}{\pi}\frac{r}{R_2}w_*^2 \tag{3.10}$$

$$W_1^2 R_1 = W_2^2 R_2$$

therefore (3.11)

$$a = \frac{W_1}{W_2} = \frac{W_2 R_2}{W_1 R_1} = \tau^2 = 2.618$$

Feigenbaum's constant a equal to 2.618 at Eq. (3.11) above represents the following dynamical and statistical properties of the period doubling route to eddy growth process

- The relative magnitude of the large eddy domain W_2R_2 with respect to the enclosed energizing primary turbulent eddy domain W_1R_1 and is equal to τ^2 (=2.618) and is inherently negative.
- A quantitative measure of the number of units of fractal domain W_1R_1 embedded in the large eddy domain.
- The non-dimensional total outward fractional mass dispersion rate and is inherently negative and $2a^2$ represents the bidirectional fractional energy flux into the large eddy environment.
- The inverse of the *Feigenbaum*'s constant a represents the steady state fractional volume dilution of large eddy by turbulent eddy mixing which is equal to the *von karman*'s constant k in the logarithmic law of the wall (Eq. 3.4).

3.4.2 *Feigenbaum's Constant* d

Feigenbaum's constant d is the successive spacing intervals of $1/z$ ($z = R/r$), i.e. spacing intervals of r/R in Eq. 3.1. The successive values of length scale ratio z for eddy growth process are $\frac{R_1}{r}, \frac{R_2}{r}, \frac{R_3}{r}, \ldots\ldots\ldots$

Therefore

$$d = \frac{r}{R_1} \times \frac{R_2}{r} = \frac{R_2}{R_1} \tag{3.12}$$

It is shown in the following that *Feigenbaum*'s constant d represents the relative spin angular moments of the total volume of large and turbulent eddies

From Eq. (3.10)

$$\frac{R_1}{R_2} = \frac{W_2^2}{W_1^2}$$

$$\left(\frac{R_1}{R_2}\right)^2 = \frac{W_2^4}{W_1^4}$$

$$\left(\frac{R_1}{R_2}\right)^2 \left(\frac{R_2}{R_1}\right)^3 = \frac{W_2^4}{W_1^4} \left(\frac{R_2}{R_1}\right)^3$$

Therefore

$$d = \frac{R_2}{R_1} = \frac{W_2^4}{W_1^4} \left(\frac{R_2}{R_1}\right)^3 \tag{3.13}$$

In Eq. (3.13) $\frac{W_2^4}{W_1^4}\left(\frac{R_2}{R_1}\right)^3$ represents the ratio of spin angular moments of total volume of large and turbulent eddies.

From Eq. (3.10) $W_1^2 R_1 = W_2^2 R_2$.

$$d = \frac{R_2}{R_1} = \frac{W_2^4}{W_1^4} \left(\frac{R_2}{R_1}\right)^3 = \frac{W_2^2 R_2}{W_1^2 R_1} \frac{W_2^2 R_2^2}{W_1^2 R_1^2} = \frac{W_2^2 R_2^2}{W_1^2 R_1^2} \tag{3.14}$$

Feigenbaum's constant $d = \frac{R_2}{R_1} = \frac{W_2^4}{W_1^4}\left(\frac{R_2}{R_1}\right)^3 = \frac{W_2^2 R_2^2}{W_1^2 R_1^2}$ at Eq. (3.14) represents the following dynamical and statistical properties of the period doubling route to eddy growth process

- The dynamic relative spin angular momentum $\frac{W_2^4}{W_1^4}\left(\frac{R_2}{R_1}\right)^3$ of the total mass of the large and turbulent eddies generates the outward energy flux $\frac{W_2^2 R_2^2}{W_1^2 R_1^2}$ in fractal structures in the large eddy environment.
- The statistical relative intermittency of occurrence of fractal structures $\frac{W_2^4}{W_1^4}\left(\frac{R_2}{R_1}\right)^3$ in the large eddy environment contributes to the relative variance $\frac{W_2^2 R_2^2}{W_1^2 R_1^2}$ of the fractal structures.
- The *Feigenbaum*'s constant d represents the scale ratio R_2/R_1 and the inverse of the Feigenbaum's constant d equal to R_1/R_2 represents the probability $(Prob)_1$ of occurrence of length scale R_1 in the total fluctuation length domain R_2 for the first eddy growth step as given in the following

  From Eq. 3.14

$$(Prob)_1 = \frac{R_1}{R_2} = \frac{1}{d} = \frac{W_1^2 R_1^2}{W_2^2 R_2^2} = \tau^{-4}$$

- In general for the nth eddy growth step, the probability $(Prob)_n$ of occurrence of length scale R_1 in the total fluctuation length domain R_n is given in terms of the normalized deviation t as (from Eq. 3.14)

$$(Prob)_n = \frac{R_1}{R_n} = \frac{W_1^2 R_1^2}{W_n^2 R_n^2} = \tau^{-4n} = \tau^{-4t}$$

- The above equation for probability $(Prob)_n$ also represents, for the nth eddy growth step, the following statistical and dynamical quantities of the growing large eddy with respect to the initial perturbation domain:

  a. probability of occurrence of fractal domain in either positive or negative direction
  b. The normalized variance spectrum $\frac{W_1^2 R_1^2}{W_n^2 R_n^2}$ expressed as the fractional probability of occurrence of any component eddy variance with respect to the nth eddy variance which represents the cumulative sum of component eddy variances.
  c. The probability of occurrence of fractal domains $\frac{R_1}{R_n}$ also represents the normalized variance spectrum $\frac{W_1^2 R_1^2}{W_n^2 R_n^2}$, i.e. the additive amplitudes of eddies when squared represent the probability densities, a result observed in quantum systems such as the electron or photon (Maddox 1988; See Sect. 3.5 for a brief review of current status of quantum mechanics). Therefore period doubling route to chaos eddy growth processes exhibit quantumlike chaos.

3.4.3 Feigenbaum's Constants' Quantification of Universal Period Doubling Route to Chaos Eddy Growth Processes

The universal algorithm for period doubling route to chaos eddy growth at Eq. 3.1 may be written in terms of Feigenbaum's constants ***a*** and ***d*** as follows.

$$\frac{2}{\pi}\frac{r}{R}w_*^2 = W^2$$

$$\frac{2}{\pi}\left(\frac{WR}{w_* r}\right)^2 \frac{r}{R}w_*^2 = W^2\left(\frac{WR}{w_* r}\right)^2$$

$$\frac{2}{\pi}\frac{W^2 R^2}{w_*^2 r^2} = \frac{W^4 R^3}{w_*^4 r^3}$$

For adjacent period doublings the above equation may be written as

$$\frac{2}{\pi}\frac{W_2^2R_2^2}{W_1^2R_1^2} = \frac{W_2^4R_2^3}{W_1^4R_1^3} \tag{3.15}$$

From Eq. 3.11 *Feigenbaum*'s constant $a = \frac{W_2R_2}{W_1R_1}$

From Eq. 3.13 *Feigenbaum*'s constant $d = \frac{W_2^4}{W_1^4}\left(\frac{R_2}{R_1}\right)^3$

Therefore Eq. 3.15 may be written as

$$\frac{2a^2}{\pi} = d \tag{3.16}$$

Thealgorithm for universal period doubling route to chaos eddy growth process at Eq. 3.1 may be written as

$$2a^2 = \pi d \tag{3.17}$$

A complete large eddy circulation of the quasiperiodic Penrose tiling pattern (Fig. 3.1) consists of angular rotation equal to π radians simultaneously on either side (clockwise and counterclockwise). The outward propagating spin angular momentum ***d*** of the fractal structures in the large eddy is associated with bidirectional energy flux $2a^2/\pi$ per unit radian in the large eddy volume.

Numerical computations by Delbourgo (1986) give the relation, $2a^2 = 3d$ which is almost identical to the model predicted Eq. 3.17.

3.4.4 *The Numerical Values of Feigenbaum's Constants* a *and* d

Eddy growth occurs in two stages (i) primary dominant (identifiable) eddy growth of radius r and circulation speed w_* consisting of 10 unit length step growths (Eq. 3.1) (ii) subsequent formation of large eddies as envelopes enclosing primary dominant eddies at successive length steps R_1 and R_2 with corresponding eddy circulation speeds W_1 and W_2 (Eq. 3.10). The growth trajectory of large eddy circulations trace the quasi-periodic Penrose tiling pattern (Fig. 3.1) such that the successive values of large eddy radii R and corresponding circulation speeds W follow the Fibonacci mathematical number series, i.e. the ratio of successive circulation speeds W and also corresponding eddy radii R is equal to the golden mean τ (≈1.618).

3.4.4.1 Primary Dominant Eddy Growth: Numerical Value of Feigenbaum's Constant a

Starting from unit length step, primary dominant eddy forms after 10 length step growths. Therefore $R_1 = 1$ and $R_2 = 11$ and the scale ratio $z = R_2/R_1 = 11$

Feigenbaum's constant ***a*** is the successive spacing intervals of W in Eq. 3.1

$$a = \frac{w}{W} = \frac{W_1}{W_2}$$

From Eq. 3.11

$$\frac{W_1}{W_2} = \frac{W_2 R_2}{W_1 R_1}$$

From Eq. 3.1

$$W_2^2 = \tfrac{2}{\pi}\tfrac{R_1}{R_2} W_1^2$$

$$\frac{W_2^2 R_2^2}{W_1^2 R_1^2} = \frac{2}{\pi}\frac{R_2}{R_1}$$

Feigenbaum's constant ***a*** is equal to

$$\frac{W_1}{W_2} = \frac{W_2 R_2}{W_1 R_1} = \sqrt{\frac{2z}{\pi}} = \sqrt{2 \times \frac{7}{22} \times 11} \approx 2.646 \quad (3.18)$$

3.4.4.2 Primary Dominant Eddy Growth: Numerical Value of Feigenbaum's Constant *d*

From Eq. (3.17)

$$d = \frac{2a^2}{\pi} \quad (3.19)$$

Substituting the numerical value for Feigenbaum's constant ***a*** equal to 2.646 in Eq. (3.18) for primary dominant eddy growth, the corresponding numerical value of Feigenbaum's constant ***d*** is obtained as

$$d = \frac{2a^2}{\pi} = \frac{2}{\pi} \times \frac{2z}{\pi} = \frac{4 \times 11}{\pi^2} \approx 4.455$$

Table 3.1 Feigenbaum's constants: comparison with model predicted values

Experiment	Feigenbaum's constant ***a***	Feigenbaum's constant ***d***
Primary eddy growth	−2.646	4.455
Large eddy growth	−2.618	4.361
Feigenbaum's result	−2.5029	4.669

The computed values of *Feigenbaum*'s universal constants for primary eddy growth and for the subsequent large eddy growth by period doubling is given in the following Table 3.1.

3.4.4.3 Subsequent Large Eddy Growth: Numerical Value of Feigenbaum's Constant *a*

(i) For large eddy growth from primary dominant eddy circulations Eq. 3.11 gives Feigenbaum's constant $a = \frac{W_2 R_2}{W_1 R_1} = \tau^2 = 2.618$
(ii) For large eddy growth from primary dominant eddy circulations the value of Feigenbaum's constant ***d*** is computed from Eq. 3.19 as follows

From Eq. 3.19

$$d = \frac{2a^2}{\pi} = \frac{2 \times \tau^4}{\pi} = 4.361$$

3.4.4.4 Feigenbaum's Constants: Comparison with Model Predicted Values

Feigenbaum's universal constants ***a*** = 2.503 and ***d*** = 4.6692 (Feigenbaum 1980) have been determined by numerical computations at period doublings n, $n + 1$ and $n + 2$ where n is large. At large n, computational difficulties in resolution of adjacent period doublings impose a limit to the accuracy with which ***a*** and ***d*** can be estimated.

The Feigenbaum's constants ***a*** and ***d*** computed from the universal algorithm $2a^2 = \pi d$ refer to an infinitesimally small value for w_* and r, i.e. an infinitely large number of period doublings. The model predicted and computed ***a*** and ***d*** are therefore not identical.

Linage et al. (2006) show that the Fibonacci sequence will be found in all dynamical systems exhibiting the period doubling route to chaos.

3.4.4.5 Eddy Dynamics and Universal Feigenbaum's Constants

In terms of eddy dynamics, the above Eq. 3.17, namely, $2a^2 = \pi d$ states that during each length step growth, the energy flux into the environment equal to $2\boldsymbol{a}^2$ contributes to generate relative spin angular momentum equal to $\pi\boldsymbol{d}$ of the growing fractal structures. Each length step growth is therefore associated with a factor of $2\boldsymbol{a}^2$ equal to $2\tau^4$ ($\cong$13.708203) increase in energy flux in the associated fractal domain. Ten such length step growths results in the formation of robust (self-sustaining) dominant bidirectional large eddy circulation $OR_OR_1R_2R_3R_4R_5$ (Fig. 3.1) associated with a factor of $20\boldsymbol{a}^2$ equal to 137.08203 increase in eddy energy flux. This non-dimensional constant factor characterizing successive dominant eddy energy increments is analogous to the *fine structure* constant $\propto^{-1}$ observed in atomic spectra, where the spacing (energy) intervals between adjacent spectral lines is proportional to the non-dimensional *fine structure* constant equal to approximately 1/137. Further, the probability of *n*th length step eddy growth is given by $\boldsymbol{a}^{-2n}$ ($\cong 6.8541^{-n}$) while the associated increase in eddy energy flux into the environment is equal to $\boldsymbol{a}^{2n}$ ($\cong 6.8541^n$). Extreme events occur for large number of length step growths *n* with small probability of occurrence and are associated with large energy release in the fractal domain. Each length step growth is associated with one-tenth of *fine structure constant* energy increment equal to $2\boldsymbol{a}^2$ ($\propto^{-1/10} \cong 13.7082$) for bidirectional eddy circulation, or equal to one-twentieth of *fine structure constant* energy increment equal to $\boldsymbol{a}^2$ ($\propto^{-1/20} \cong 6.8541$) in any one direction, i.e., positive or negative. The energy increase between two successive eddy length step growths may be expressed as a function of $(\boldsymbol{a}^2)^2$, i.e., proportional to the square of the *fine structure constant* $\propto^{-1}$. In the spectra of many atoms, what appears with coarse observations to be a single spectral line proves, with finer observation, to be a group of two or more closely spaced lines. The spacing of these fine structure lines relative to the coarse spacing in the spectrum is proportional to the square of *fine structure constant*, for which reason this combination is called the *fine structure constant*. We now know that the significance of the *fine structure constant* goes beyond atomic spectra (Ford 1968).

3.5 Basic Concepts in Quantum Mechanics: Current Status

Quantum theory is based on a clear mathematical apparatus, has enormous significance for the natural sciences, enjoys phenomenal predictive success, and plays a critical role in modern technological developments. Yet, nearly 90 years after the theory's development, there is still no consensus in the scientific community regarding the interpretation of the theory's foundational building blocks (Schlosshauer et al. 2013a, b).

Vedral (2013) states that Quantum physics departs from classical physics in two key aspects, namely (i) most fundamental events in the microscopic world are genuinely random and (ii) the existence of correlations—how measurements on one system give results related to measurements on another—which are independent of space and time, and connect objects into a network of interdependent entities. Both these properties of quantum physics, however, have since been confirmed in numerous experiments. Quantum randomness is now being used in cryptography to improve the secrecy of communications and quantum entanglement provides the basis for many of the fast-developing quantum technologies.

Quantum mechanics is an outstandingly successful description of nature, underpinning fields from biology through chemistry to physics. The central tool for describing quantum systems is the quantum wavefunction ψ. Yet it is still unclear what the wavefunction actually is: does it merely represent our limited knowledge of a system, or is it in direct correspondence to reality (Ringbauer et al. 2015).

Classical mechanics (Newtonian mechanics) can explain macroscopic phenomena while quantum mechanics is used to explain microscopic dynamics of quantum systems such as the electron or photon. In Newtonian mechanics objects are solid particles and the laws are written in terms of precisely defined particle trajectories. Quantum mechanics is a collection of postulates based on a large number of observations. The following observed characteristics of microscopic scale quantum systems such as the electron or photon do not have satisfactory real world physical explanations. (i) Quantum particles can act as both waves (eddies) and particles and possess wave–particle duality and as a result the position and momentum of the particle are indeterminate, i.e., can be given in terms of probabilities only. (ii) The square of wave amplitude gives the probability of occurrence of the quantum system at that location. (iii) Any measurement on quantum system disturbs the state of the system in an unpredictable manner. (iii) Energy of quantum particles is quantised. (iv) The separated parts of a quantum system responds as a unified whole to local perturbations manifested as non-local connection.

In the following (Sects. 3.5.1–3.5.3) it is shown that the general systems theory model for fractal fluctuations in dynamical systems (Selvam 2007) predicts that the macroscale real world phenomena of weather systems exhibit the above listed quantumlike behaviour as a natural consequence of intrinsic eddy (wave) continuum structure of atmospheric flows which functions as a unified whole communicating network with long-range space-time correlations.

Palmer (2005, 2009, 2014) has investigated the role of fractals in quantum theory (Sect. 3.5.1). The string-like energy flow pattern of atmospheric eddy continuum is similar to string theory for quantum phenomena put forth by 't Hooft (2014) (Sect. 3.5.2). Bush (2015) (Sect. 3.5.3) suggests fluid mechanical concepts may help explain quantum mechanical phenomena.

3.5.1 Fractals and Quantum Theory

The role of fractals in fundamental physics was investigated by Palmer (2005, 2009, 2014). Palmer (2005) discusses possible linkages between nonlinear paradigms developed to understand meteorological predictability and the deepest conceptual problems of quantum theory. Application of nonlinear meteorological thinking may indeed provide fresh insights on the foundational problems of quantum theory. Quantum theory is the most successfully tested, yet least well understood, of all physical theories (see, e.g., Penrose 1989). Einstein's dissatisfaction with quantum theory is well known; the two key reasons for such dissatisfaction: indeterminacy and nonlocal causality. The notion that fractals may play a role in fundamental physics is not itself new. However, while earlier studies have focused on the concept that space–time itself may be fractal (Ord 1983; Nottale and Schneider 1984; El Naschie 2004), Palmer considers the ontological significance of fractal geometry in state space.

In the 1960s, the introduction of global space–time geometric and topological methods, transformed our understanding of classical gravitational physics (Penrose 1965). It is proposed that the introduction of global geometric and topological methods in state space may similarly transform our understanding of quantum gravitational physics. Combining these rather different forms of geometry may provide the missing element needed to advance the search for a unified theory of physics (Palmer 2009). Arguments over quantum theory have raged since the 1920s. The mathematics of fractals may help to understand the long-standing puzzles of quantum theory. Very few scientists working on fundamental physics have explored how fractals might be incorporated into the theory, even though they are commonplace in other parts of physics (Buchanan 2009). Palmer (2014) states that there were three great revolutions in 20th Century theoretical physics: relativity theory, quantum theory and chaos theory. It is proposed that the three great revolutions of 20th Century physics can be unified if the universe is considered a deterministic dynamical system evolving on a fractal invariant set. Each has had a profound impact on the development of science, and yet their domains of impact remain quite distinct. Despite over a half century of intense research, there is still no consensus on how to combine quantum theory and general relativity theory into a supposed 'quantum theory of gravity', nor even a consensus about what such a notion means physically. Moreover, the unpredictability of nonlinear chaotic systems is generally considered quite unrelated to the indeterminism of quantum measurement. Development of new ideas based on fractal invariant set may lead to some unification of these three revolutions. These ideas evolve around Einstein's great insight that geometry provides the ultimate expression of the laws of physics (Palmer 2014).

Buchanan (2014) summarizes Palmer's work on fractals and quantum mechanics as follows. Most physicists now take the view that quantum physics is irreducibly non-deterministic and that nature is fundamentally ruled by chance. Palmer now believes that the theory of dynamical chaos and 'strange attractors'—the geometrical structures signifying chaos in dissipative systems—might be the key concepts required to tackle a number of fundamental physics issues, among them building a sensible theory of quantum gravity. Moreover, he suggests that a return to determinism might follow. In dynamical systems theory, the key object in the description of the long-term dynamics of a dissipative system is its attractor—an invariant set that almost all system trajectories approach asymptotically. In a chaotic system, this is a strange attractor with fractal or multifractal geometry. In the famous Lorenz system, for example, the attractor looks crudely like two intersecting surfaces that resemble a butterfly. On closer inspection it turns out to be an infinitely intricate set of nested surfaces—a fractal set of non-integer dimension. Palmer makes a conjecture—but a natural one—that the dynamics of the stuff of our Universe may be similarly described as approaching some invariant attractor. Palmer refers to this as the 'invariant set postulate'—that the Universe is evolving causally and deterministically on (or very close to) some measure zero, fractal invariant set. This would lead to an attractor. The world of deterministic dynamics, he suggests, already has enough weirdness in it to account for everything in quantum physics—but only if we really take deterministic chaos seriously at the universal scale.

3.5.2 Quantum Mechanics and String Theory

In recent years the string field theory has been proposed for a realistic explanation of quantum mechanical laws. Itzhak and Rychkov (2014) have proposed a link between string field theory and quantum mechanics as the basis of all physics. The essential ingredient is the assumption that all matter is made up of strings and that the only possible interaction is joining/splitting as specified in their version of string field theory. Physicists have long sought to unite quantum mechanics and general relativity, and to explain why both work in their respective domains. First proposed in the 1970s, string theory resolved inconsistencies of quantum gravity and suggested that the fundamental unit of matter was a tiny string, not a point, and that the only possible interactions of matter are strings either joining or splitting. At present, no single set of rules can be used to explain all of the physical interactions that occur in the observable universe (Perkins 2014). 't Hooft (2014) suggests that superstrings may well form exactly the right mathematical system that can explain how quantum mechanics can be linked to a deterministic picture of our world.

3.5.3 Fluid Mechanics and Quantum Mechanics

Bush (2015) states that thinking of space and time as a liquid might help reconcile quantum mechanics and relativity. If space-time is like a liquid—a concept some physicists say could help resolve a confounding disagreement between two dominant theories (quantum mechanics and relativity) in physics—it must be a very special liquid indeed. A recent study compared astrophysical observations with predictions based on the notion of fluid space-time, and found the idea only works if space-time is incredibly smooth and freely flowing—in other words, a superfluid (Moskowitz 2014).

Bohmian mechanics is a hydrodynamic formulation of the quantum theory and helps to understand the role of the phase as the mechanism responsible for the dynamical evolution of quantum systems. The Bohmian formulation helps to explain in a natural way aspects of the quantum theory typically regarded as paradoxical. They show that a proper understanding of quantum phase dynamics immediately removes any trace of unnecessary artificial wave–particle arguments. This representation is similar to classical statistical treatments, where physically meaningful quantities arise from ensembles rather than from single trajectories (Sanz 2015).

3.5.4 General Systems Theory for Fractal Space-Time Fluctuations and Quantum-like Chaos in Atmospheric Flows

Atmospheric flows, a representative example of turbulent fluid flows, exhibit long-range spatiotemporal correlations manifested as the fractal geometry to the global cloud cover pattern concomitant with inverse power-law form for spectra of temporal fluctuations. Such non-local connections are ubiquitous to dynamical systems in nature and are identified as signatures of self-organized criticality (Bak et al. 1988). Applications of self-similarity and self-organized criticality in atmospheric sciences are currently being investigated (Dessai and Walters 2000; Yano et al. 2012; Craig and Mack 2013; Stechmann and Neelin 2014). A cell dynamical system model developed for atmospheric flows shows that the observed long-range spatiotemporal correlations are intrinsic to quantum-like mechanics governing fluid flows (Selvam 1990, 2007). The model concepts are independent of the exact details such as the chemical, physical, physiological and other properties of the dynamical system and therefore provide a general systems theory applicable to all real world and computed dynamical systems in nature.

The model is based on the concept that spatial integration of enclosed small-scale fluctuations results in the formation of large eddy circulations. The model predicts the following: (a) The flow structure consists of an overall logarithmic spiral trajectory with the quasiperiodic Penrose tiling pattern (Fig. 3.1) for the internal structure such that the ratio of successive eddy radii/circulation speeds is equal to the golden ratio τ ($\approx$1.618). (b) Conventional power spectrum analysis will resolve such spiral trajectory as a continuum of eddies with progressive increase in phase. (c) Increments in phase angle are concomitant with increase in period length and also represents the variance, a characteristic of quantum systems identified as 'Berry's phase'. (d) The quantum mechanical constants 'fine structure constant' and 'ratio of proton mass to electron mass' which are pure numbers and obtained by experimental observations only, are now derived in terms of the golden ratio. (g) Atmospheric flow structure follows Kepler's third law of planetary motion. Therefore Newton's inverse square law for gravitation applies to eddy masses also. The centripetal accelerations representing inertial masses (of eddies) are equivalent to gravitational masses. Fractal structure to the space-time continuum can be visualized as a nested continuum of vortex (eddy) circulations whose inertial masses obey Newton's inverse square law of gravitation. The model concepts are equivalent to a superstring model for subatomic dynamics, which incorporates gravitational forces. The fractal structure to space-time and also fractalisation of microspace is the origin of gravity (Argyris and Ciubotariu 1997; El Naschie 2004).

3.5.5 Model Predictions and the Interpretation of Quantum Mechanical Laws

The model predictions (Selvam 1990, 2007, 2015) and the interpretation of quantum mechanical laws as applied to macroscale fluid flows are described. It is shown that the apparent paradoxes of quantum mechanics are physically consistent in the context of atmospheric flows.

Almost 100 years ago physicists Werner Heisenberg, Max Born und Erwin Schrödinger created a new field of physics: quantum mechanics. Objects of the quantum world—according to quantum theory—no longer move along a single well-defined path. Rather, they can simultaneously take different paths and end up at different places at once. Physicists speak of quantum superposition of different paths (Phys.org 2015).

Researchers of the University of Bonn have shown that Caesium atoms do not follow well-defined paths. Their first experiment shows that Caesium atoms can indeed take two paths at the same time (Robens et al. 2015).

According to quantum theory, there is no intrinsic upper limit on size or complexity of a physical system above which quantum effects no longer occur (Aspelmeyer and Zeilinger 2008). Classical physics has equal status with quantum mechanics; it is but a useful approximation of a world that is quantum at all scales. Although quantum effects may be harder to see in the macroworld, the reason has nothing to do with size per se but with the way that quantum systems interact with one another. Until the past decade, experimentalists had not confirmed that quantum behaviour persists on a macroscopic scale. Today, however, they routinely do. These effects are more pervasive than anyone ever suspected (Vedral 2011). Nanotechnologists at the University of Twente research institute MESA+, The Netherlands (Strambini et al. 2016) have discovered a new fundamental property of electrical currents in very small metal circuits. They show how electrons can spread out over the circuit like waves and cause interference effects at places where no electrical current is driven. The geometry of the circuit plays a key role in this so called nonlocal effect. The interference is a direct consequence of the quantum mechanical wave character of electrons and the specific geometry of the circuit.

3.5.5.1 Probability and Amplitude Square: Probability of Weather System

Atmospheric flows trace an overall logarithmic spiral trajectory $OR_OR_1R_2R_3R_4R_5$ simultaneously in clockwise and anti-clockwise directions with the quasi-periodic Penrose tiling pattern (Steinhardt 1997) for the internal structure shown in Fig. 3.1.

The spiral flow structure can be visualized as an eddy continuum generated by successive length step growths OR_O, OR_1, OR_2, OR_3,….respectively equal to R_1, R_2 R_3 … which follow Fibonacci mathematical series such that $R_{n+1} = R_n + R_{n-1}$ and $\frac{R_{n+1}}{R_n} = \tau$ where τ is the golden mean equal to $(1 + \sqrt{5})/2$ ($\approx$1.618). Considering a normalized length step equal to 1 for the last stage of eddy growth, the successively decreasing radial length steps can be expressed as 1, $1/\tau$, $1/\tau^2$, $1/\tau^3$, …… The normalized eddy continuum comprises of fluctuation length scales 1, $1/\tau$, $1/\tau^2$, …….. The probability of occurrence is equal to $1/\tau$ and $1/\tau^2$ respectively for eddy length scale $1/\tau$ in any one or both rotational (clockwise and anti-clockwise) directions. Eddy fluctuation length of amplitude $1/\tau$, has a probability of occurrence equal to $1/\tau^2$ in both rotational directions, i.e. the square of eddy amplitude represents the probability of occurrence in the eddy continuum. Similar result is observed in the subatomic dynamics of quantum systems which are visualized to consist of the superimposition of eddy fluctuations in wave trains (eddy continuum). Boeyens and Thackeray (2014) discuss the remarkable cosmic occurrence of the golden ratio τ with reference to space-time, relativity and quantum mechanics. Lindner et al. (2015) show that the unprecedented light curves of the Kepler space telescope document how the brightness of some stars pulsates at primary and secondary frequencies whose ratios are near the golden mean, the most irrational number. Coldea et al. (2010) discovered Golden mean (golden ratio) τ in quantum world, in the hidden symmetry observed for the first time in solid state matter. Heyrovska and Narayan (2005), Heyrovska (2009, 2013) has found that the golden ratio in the creations of nature arises in the architecture of atoms and ions. Selvam (2013, 2014, 2015) has shown that the amplitude as well as the variance of fractal fluctuations follow the same inverse power-law distribution signifying quantumlike dynamics in atmospheric flows.

3.5.5.2 Non-local Connection in Weather Systems

Atmospheric flows: weather systems in updraft regions and weather dissipation in adjacent down draft regions.

Nonlocal connections are intrinsic to the space-time geometry of quasiperiodic Penrose titling pattern traced by the atmospheric flow pattern visualized as an extended object. The instantaneous nonlocal connections in the string-like energy flow patterns which represent extended objects such as clouds in atmospheric flows can be visualized as shown in Fig. 3.2. The circulation flow pattern with centre O and radius OU (or OD) in Fig. 3.2 represents an eddy. In the medium of propagation, namely, atmosphere (air) in this case, upward motion U represents convection and cloud formation in association simultaneously with cloud dissipation in downward motion D. There is an instantaneous nonlocal connection between the phases of the particles at U and D. The same concept can be applied to an extended object (Fig. 3.3) such as a row of clouds represented by the wave function ψ which results from the superimposition of a continuum of eddies. The weather system at Fig. 3.3 is an extended object and the severity of the associated weather at any location is given by the square of eddy amplitude ψ at the location. Since the eddy amplitude ψ is the equal to the superimposition of component eddies with two-way ordered energy flow between the scales, any perturbation at one location is transmitted to all the component eddies. Large eddy circulations are integrated means of enclosed primary small scale circulations. For a fixed primary small–scale eddy the successively larger eddy circulations are in multiples of the primary small-scale eddy energy and therefore large eddy energy may be considered to be quantized.

3.5.5.3 Atmospheric Teleconnections: A Signature of Quantumlike Global Scale Non-local Connections

Although the periodic climatic phenomenon known as El Niño originates over the Pacific, it can affect climates all over the globe, creating warmer winters in the US, droughts in Indonesia or cold spells in northern Europe. It is an example of a climate "teleconnection," a link that intertwines distant geographic regions. Identifying possible teleconnections and what causes them is essential for accurate climate modelling (Rini 2015). Zhou et al. (2015) model the climate system as a network in which geographic locations are represented by nodes connected by links whose "weights" depend on correlations between the nodes. They found over 200 long-distance teleconnections and suggest that a large fraction of them are due to Rossby waves—high-altitude airwaves that redistribute energy across the planet.

Generally speaking, most teleconnections are caused by the transport of energy and propagation of waves (Liu and Alexander 2007). However, the mechanisms of specific teleconnections are still not well understood, since these known patterns describe the mature phase of the variability instead of the developing phase (Nigam and Baxter 2015). Therefore, deeper investigations on the evolutionary process in the spatiotemporal domain are needed.

3.5.5.4 Non-local Connection in Quantum Systems

The phenomenon known as nonlocality or "action at a distance" characterize quantum systems. Experiments in quantum optics show that two distant events can influence each other instantaneously. Nonlocal connections in quantum systems apparently violate the fundamental theoretical law in modern physics that signal transmission cannot exceed the speed of light. The distinction between locality and nonlocality is related to the concept of a trajectory of a single point object (Chiao et al. 1993).

An important goal for researchers in this field has been to confirm that such nonlocal correlations really exist in nature (Phys.org. 2014; Tura et al. 2014). Perhaps nature is indeed more nonlocal than is described in quantum mechanics says, but we haven't yet observed such a situation experimentally (Popescu 2014). Experiments reported by Hensen et al. (2015) definitively demonstrate that the quantum world violates locality, the principle that events sufficiently separated in space-time must be independent. No one knows how to describe the separation between the bizarre quantum world where entanglement exists and the everyday world where nature appears to operate via easy-to-spot causes and effects. Creating entanglement on a larger scale may help clarify this mysterious division (Sanders 2010).

Aspect (2015) has cited three recent independent groups who have done stringent experimental tests to show that non-locality, i.e. action at a distance is real for quantum systems (Hensen et al. 2015; Giustina et al. 2015; Shalm et al. 2015).

In Sect. 2.5.5.2 above it is shown that non-local connections are intrinsic to macroscale atmospheric eddy flow circulations.

References

Argyris, J., Ciubotariu, C.: On El Naschie's complex time and gravitation. Chaos, Solitons Fractals **8**, 743–751 (1997)

Aspect, A.: Viewpoint: closing the door on Einstein and Bohr's quantum debate. Physics **8**, 123 (4 pages) (2015)

Aspelmeyer, M., Zeilinger, A.: Feature: a quantum renaissance. Phys. World 22–28 (2008)

Bak, P., Chen, K.: The physics of fractals. Physica D **38**, 5–12 (1989)

Bak, P., Chen, K.: Self-organized criticality. Sci. Am. **264**(1), 46–53 (1991)

Bak, P., Tang, C., Wiesenfeld, K.: Self-organized criticality: an explanation of 1/*f* noise. Phys. Rev. Lett. **59**, 381–384 (1987)

Bak, P.C., Tang, C., Wiesenfeld, K.: Self-organized criticality. Phys. Rev. A **38**, 364–374 (1988)

Berry, M.V.: The geometric phase. Sci. Amer. **259**(6), 26–34 (1988)

Boeyens, J.C.A., Thackeray, J.F.: Number theory and the unity of science. S. Afr. J. Sci. **110** (11/12), 5–6 (2014)

Brown, J.: Where two worlds meet. New Sci. **150**(2030), 26–30 (1996)

Buchanan, M.: One law to rule them all. New Sci. **156**(2107), 30–35 (1997)

Buchanan, M.: Fractal reality. New Sci. **201**(2701), 37–39 (2009)

Buchanan, M.: Does not compute? Nat. Phys. **10**(6), 404 (2014)

Bush, J.W.M.: Pilot-wave hydrodynamics. Annu. Rev. Fluid Mech. **47**, 269–292 (2015)

Chiao, R.Y., Kwiat, P.G., Steinberg, A.M.: Faster than light. Sci. Am. **269**(2), 52–60 (1993)

Coldea, R., Tennant, D.A., Wheeler, E.M., Wawrzynska, E., Prabhakaran, D., Telling, M., Habicht, K., Smeibidl, P., Kiefer, K.: Quantum criticality in an ising chain: experimental evidence for emergent E8 symmetry. Science **327**(5962), 177–180 (2010)

Craig, G.C., Mack, J.M.: A coarsening model for self-organization of tropical convection. J. Geophys. Res. Atmos. **118**, 8761–8769 (2013)

Delbourgo, R.: Universal facets of chaotic processes. Asia-Pacific Physics News **1**, 7–11 (1986)

Dessai, S., Walter, M.E.: Self-organized criticality and the atmospheric sciences: selected review, new findings and future directions (2000). http://www.esig.ucar.edu/extremes/papers/walter.PDF

Ecke, R.E.: Chaos, patterns, coherent structures, and turbulence: reflections on nonlinear science. Chaos: An Interdisciplinary Journal of Nonlinear Science **25**, 097605(8 pages) (2015)

El Naschie, M.S.: A review of E-infinity theory and the mass spectrum of high energy particle physics. Chaos, Solitons Fractals **19**, 209–236 (2004)

Feigenbaum, M.J.: Universal behavior in nonlinear systems. Los Alamos Sci. **1**, 4–27 (1980)

Ford, K.W.: Basic physics. Blaisdell Publishing Company, Waltham, Massachusetts, USA (1968)

Giustina, M., Versteegh, M.A.M., Wengerowsky, S., et al.: 2015: Significant-loophole-free test of bell's theorem with entangled photons. Phys. Rev. Lett. **115**, 250401(7 pages)

Gleick, J.: Chaos: making a new science. Viking, New York (1987)

Grossing, G.: Quantum systems as order out of chaos phenomena. Il Nuovo Cimento **103B**, 497–510 (1989)

Hensen, B., Bernien, H., Dréau, A.E., et al.: Loophole-free Bell inequality violation using electron spins separated by 1.3 kilometers. Nature **526**, 682–686 (2015)

Heyrovska, R.: Golden ratio based fine structure constant and rydberg constant for hydrogen spectra. Int. J. Sci. **2**, 28–31 (2013)

Heyrovska, R., Narayan, S.: Fine-structure constant, anomalous magnetic moment, relativity factor and the golden ratio that divides the Bohr radius (2005) http://arxiv.org/ftp/physics/papers/0509/0509207.pdf

Heyrovska, R.: The golden ratio in the creations of nature arises in the architecture of atoms and ions. In: Şener, B. (ed.) Innovations in Chemical Biology, pp. 133–139. Springer (2009)

Iovane, G., Chinnici, M., Tortoriello, F.S.: Multifractals and El Naschie E-infinity Cantorian space–time. Chaos, Solitons Fractals **35**(4), 645–658 (2008)

Itzhak, B., Rychkov, D.: Background independent string field theory (2014) arXiv:1407.4699v2

Jean, R.V.: Phyllotaxis: A Systemic Study in Plant Morphogenesis. Cambridge University Press, New York (1994)

Kaku, M.: Into the eleventh dimension. New Sci. **153**(2065), 32–36 (1997)

Kolmogorov, A.N.: The local structure of turbulence in incompressible viscous fluid for very large Reynolds numbers. C. R. Acad. Sci. URSS **30**, 301–305 (1941)

Konar, S.: A divine surprise: the golden mean and the round-off error. Resonance **11**(4), 91–99 (2006)

Li, M., Zhao, W.: Golden ratio phenomenon of random data obeying von Karman spectrum. Math. Probl. Eng. **2013**, 1–6 (2013a). doi:10.1155/2013/130258

Li, M., Zhao, W.: Essay on Kolmogorov law of minus 5 over 3 viewed with golden ratio. Adv. High Energy Phys. **2013**, 1–3 (2013b). doi:10.1155/2013/680678

Linage, G., Montoya, F., Sarmiento, A., Showalter, K., Parmananda, P.: Fibonacci order in the period-doubling cascade to chaos. Phys. Lett. A **359**, 638–639 (2006)

Lindner, J.F., Kohar, V., Kia, B., Hippke, M., Learned, J.G., Ditto, W.L.: Strange nonchaotic stars. Phys. Rev. Lett. **114**, 054101 (2015)

Liu, Z., Alexander, M.: Atmospheric bridge, oceanic tunnel, and global climatic teleconnections. Rev. Geophys. **45**, RG2005 (2007). doi:10.1029/2005RG000172

Lovejoy, S., Schertzer, D.: Scale invariance, symmetries, fractal and stochastic simulations of atmospheric phenomena. Bull. Am. Meteorol. Soc. **67**, 21–32 (1986a)

Lovejoy, S., Schertzer, D.: Scale in variance in climatological temperatures and the local spectral plateau. Annates Geophysicae **86B**, 401–409 (1986b)

Maddox, J.: Licence to slang Copenhagen? Nature **332**, 581 (1988)
Maddox, J.: Can quantum theory be understood? Nature **361**(6412), 493 (1993)
Mandelbrot, B.B.: Fractals: Form, Chance and Dimension. W. H. Freeman and Co., San Francisco (1977)
Mandelbrot, B.B.: The Fractal Geometry of Nature. W. H. Freeman and Co., NY (1983)
Moskowitz, C.: If space-time were a superfluid, would it unify physics—or is the theory all wet? Sci. Am. (2014). http://www.scientificamerican.com/article/superfluid-spacetime-relativity-quantum-physics/
Moss, F., Wiesenfeld, K.: The benefits of background noise. Sci. Am. **273**(2), 66–69 (1995)
Nigam, S., Baxter, S.: Teleconnections. In: Encyclopedia of Atmospheric Sciences, 2nd ed., pp. 90–109. Elsevier Ltd, London (2015)
Nottale, L., Schneider, J.: Fractals and non-standard analysis. J. Math. Phys. **25**, 1296–1300 (1984)
Ord, G.N.: Fractal space-time: a geometric analogue of relativistic quantum mechanics. J. Phys. A **16**, 1869–1884 (1983)
Palmer, T.N.: Quantum reality, complex numbers, and the meteorological butterfly effect. Bull. Am. Meteorol. Soc. **86**, 519–530 (2005)
Palmer, T.N.: The invariant set postulate: a new geometric framework for the foundations of quantum theory and the role played by gravity. Proc. R. Soc. A **465**, 1–21 (2009)
Palmer, T.N.: Gödel and Penrose: new perspectives on determinism and causality in fundamental physics. Contemp. Phys. **55**(3), 157–178 (2014)
Penrose, R.: Gravitational collapse and space-time singularities. Phys. Rev. Lett. **14**, 57–59 (1965)
Penrose, R.: The Emperor's New Mind. Oxford University Press, UK (1989)
Perkins, R.: String field theory could be the foundation of quantum mechanics (2014). http://phys.org/news/2014-11-field-theory-foundation-quantummechanics.html
Philander, S.G.: El Nino, La Nina and the Southern Oscillation. International Geophysical Series 46, Academic Press, NY (1990)
Phys.org.: Atoms can be in two places at the same time. Retrieved 21 Jan 2015 from http://phys.org/news/2015-01-atoms.html
Phys.org.: Experimentally testing nonlocality in many-body systems (2014). http://phys.org/news/2014-06-experimentally-nonlocality-many-body.html
Popescu, S.: Nonlocality beyond quantum mechanics. Nat. Phys. **10**, 264–270 (2014)
Puthoff, H.E.: Ground state of hydrogen as a zero-point fluctuation-determined state. Phys. Rev. D **35**, 3266–3269 (1987)
Puthoff, H.E.: Gravity as a zero-point fluctuation force. Phys. Rev. A **39**, 2333–2342 (1989)
Rae, A.: Quantum-physics: Illusion or reality?. Cambridge University Press, New York (1988)
Ringbauer, M., Duffus, B., Branciard, C., Cavalcanti, E.G., White, A.G., Fedrizzi, A.: Measurements on the reality of the wavefunction. Nat. Phys. **11**, 249–254 (2015)
Rini, M.: Synopsis: finding climate teleconnection paths. Physics, Dec 2015, American Physical Society. https://physics.aps.org/synopsis-for/10.1103/PhysRevLett.115.268501
Robens, C., Alt, W., Meschede, D., Emary, C., Alberti, A.: Ideal negative measurements in quantum walks disprove theories based on classical trajectories. Phys. Rev. X **5**, 011003 (10 pages) (2015). doi:10.1103/PhysRevX.5.011003)
Sanders, L.: Everyday entanglement: physicists take quantum weirdness out of the lab. Sci. News **178**(11), 22–29 (2010)
Sanz, A.S.: Investigating puzzling aspects of the quantum theory by means of its hydrodynamic formulation. Found. Phys. **45**, 1153–1165 (2015)
Schertzer, D., Lovejoy, S.: Nonlinear Variability in Geophysics—Scaling and Fractals. Kluwer Academic press, Dordretch, Holland (1991)
Schertzer, D., Lovejoy, S.: Multifractal generation of self-organized criticality. In: Novak, M.M. (ed.) Fractals in the Natural and Applied Sciences (**A-41**), pp. 325–339. Elsevier Science B.V., North-Holland (1994)
Schlosshauer, M., Kofler, J., Zeilinger, A.: (2013a): Snapshot of foundational attitudes toward quantum mechanics. arXiv:1301.1069v1 [quant-ph] 6 Jan 2013

Schlosshauer, M., Kofler, J., Zeilinger, A.: The interpretation of quantum mechanics: from disagreement to consensus? Ann. Phys. **525**(4), A51–A54 (2013b)

Schroeder, M.: Fractals, Chaos and Power-Laws. W. H. Freeman and Co., NY (1991)

Selvam, A.M.: Deterministic chaos, fractals and quantumlike mechanics in atmospheric flows. Can. J. Phys. **68**, 831–841 (1990). http://xxx.lanl.gov/html/physics/0010046

Selvam, A.M.: Universal quantification for deterministic chaos in dynamical systems. Appl. Math. Model. **17**, 642–649 (1993). http://xxx.lanl.gov/html/physics/0008010

Selvam, A.M.: Quasicrystalline pattern formation in fluid substrates and phyllotaxis. In: Barabe, D., Jean, R.V. (eds.) Symmetry in Plants. World Scientific Series in Mathematical Biology and Medicine, vol. 4, pp. 795–809. Singapore (1998). http://xxx.lanl.gov/abs/chao-dyn/9806001

Selvam, A.M.: Cantorian fractal space-time and quantum-like chaos in neural networks of the human brain. Chaos, Solitons Fractals **10**, 25–29 (1999). http://xxx.lanl.gov/abs/chao-dyn/9809003

Selvam, A.M.: Chaotic Climate Dynamics. Luniver Press, UK (2007)

Selvam, A.M.: Fractal fluctuations and statistical normal distribution. Fractals **17**(3), 333–349 (2009). http://arxiv.org/pdf/0805.3426

Selvam, A.M.: Quantum-like chaos in prime number distribution and in turbulent fluid flows. Apeiron **8**, 29–64 (2001a). http://redshift.vif.com/JournalFiles/V08NO3PDF/V08N3SEL.PDF http://xxx.lanl.gov/html/physics/0005067

Selvam, A.M.: Signatures of quantum-like chaos in spacing intervals of non-trivial Riemann zeta zeros and in turbulent fluid flows. Apeiron **8**, 10–40 (2001b). http://redshift.vif.com/JournalFiles/V08NO4PDF/V08N4SEL.PDF http://xxx.lanl.gov/html/physics/0102028

Selvam, A.M.: Cantorian fractal space-time fluctuations in turbulent fluid flows and the kinetic theory of gases. Apeiron **9**, 1–20 (2002a). http://redshift.vif.com/JournalFiles/V09NO2PDF/V09N2sel.PDF http://xxx.lanl.gov/html/physics/9912035

Selvam, A.M.: Quantumlike chaos in the frequency distributions of the bases A, C, G, T in Drosophila DNA. Apeiron **9**, 103–148 (2002b). http://redshift.vif.com/JournalFiles/V09NO4PDF/V09N4sel.pdf http://arxiv.org/html/physics/0210068

Selvam, A.M.: Scale-free universal spectrum for atmospheric aerosol size distribution for Davos, Mauna Loa and Izana. Int. J. Bifurcat. Chaos **23**, 1350028 (13 pages) (2013) http://arxiv.org/pdf/1111.3132

Selvam, A.M.: Universal inverse power-law distribution for temperature and rainfall in the UK region. Dyn. Atmos. Oceans **66**, 138–150 (2014)

Selvam, A.M.: Rain Formation in Warm Clouds: General Systems Theory. SpringerBriefs in Meteorology. Springer (2015)

Selvam, A.M., Fadnavis, S.: Signatures of a universal spectrum for atmospheric inter-annual variability in some disparate climatic regimes. Meteorol. Atmosp. Phys. **66**, 87–112 (1998). http://xxx.lanl.gov/abs/chao-dyn/9805028

Selvam, A.M., Sen, D., Mody, S.M.S.: Critical fluctuations in daily incidence of acute myocardial infarction. Chaos, Solitons Fractals **11**, 1175–1182 (2000). http://xxx.lanl.gov/abs/chao-dyn/9810017

Selvam, A.M., Joshi, R.R.: Universal spectrum for interannual variability in COADS global air and sea surface temperatures. Int. J. Climatol. **15**, 613–624 (1995)

Selvam, A.M., Pethkar, J.S., Kulkarni, M.K.: Signatures of a universal spectrum for atmospheric interannual variability in rainfall time series over the Indian region. Int. J. Climatol. **12**, 137–152 (1992)

Selvam, A.M., Pethkar, J.S., Kulkarni, M.K., Vijayakumar, R.: Signatures of a universal spectrum for atmospheric interannual variability in COADS surface pressure time series. Int. J. Climatol. **16**, 1–11 (1996)

Shalm, L.K., Meyer-Scott, E., Christensen, B.G., et al.: Strong loophole-free test of local realism. Phys. Rev. Lett. **115**, 250402 (10 pages) (2015)

Spruch, L.: Long-range (Casimir) interactions. Science **272**, 1452–1455 (1996)

Sreenivasan, K.R.: Fractals and multifractals in turbulence. Annu. Rev. Fluid Mech. **23**, 539–600 (1991)

Stanley, H.E.: Exotic statistical physics: applications to biology, medicine, and economics. Phys. A **285**, 1–17 (2000)
Stechmann, S.N., Neelin, J.D.: First-passage-time prototypes for precipitation statistics. J. Atmos. Sci. **71**(9), 3269–3291 (2014)
Steinhardt, P.: Crazy crystals. New Sci. **153**, 32–35 (1997)
Stewart, I.: Warning—handle with care! Nature **135**, 16–17 (1992)
Stewart, I.: Sources of uncertainty in deterministic dynamics: an informal overview. Phil. Trans. R. Soc. A **369**, 4705–4729 (2011)
Strambini, E., Makarenko, K.S., Abulizi, G., de Jong, M.P., van der Wiel, W.G.: Geometric reduction of dynamical nonlocality in nanoscale quantum circuits. Sci. R. **6**, 18827 (6 pages) (2016)
't Hooft, G.: Superstrings and the foundations of quantum mechanics. Found. Phys. **44**, 463–471 (2014)
Tessier, Y., Lovejoy, S., Schertzer, D.: Universal multifractals: theory and observations for rain and clouds. J. Appl. Meteorol. **32**, 223–250 (1993)
Tura, J., Augusiak, R., Sainz, A.B., Vértesi, T., Lewenstein, M., Acín, A.: Detecting nonlocality in many-body quantum states. Science **344**(6189), 1256–1258 (2014)
Uzer, T., Farrelly, D., Milligan, J.A., Raines, P.E., Skelton, J.P.: Celestial mechanics on a microscopic scale. Science **253**, 42–48 (1991)
Vedral, V.: The curious state of quantum physics. Phys. World **26**(3), 30–34 (2013)
Vedral, V.: Living in a quantum world. Sci. Am. 38–43 (2011)
Vivaldi, F.: Periodicity and transport from round-off errors. Exp. Math. **3**(4), 303–315 (1994)
von Bertalanffy, L.: The history and status of general systems theory. Acad. Manage. J. **15**(4), General Systems Theory, 407–426 (1972)
von Karman, T.: Progress in the statistical theory of turbulence. Proc. Natl. Acad. Sci. **34**(11), 530–539 (1948)
von Kármán, Th. "Mechanische Ähnlichkeit und Turbulenz", *Nachrichten von der Gesellschaft der Wissenschaften zu Göttingen, Fachgruppe 1 (Mathematik)* **5**: 58–76(also as: "Mechanical Similitude and Turbulence", Tech. Mem. NACA, no. 611, 1931) (1930)
Wang, P., Huang, G., Wang, Z.: Analysis and application of multiple-precision computation and round-off error for nonlinear dynamical systems. Adv. Atmos. Sci. **23**(5), 758–766 (2006)
Watkins, N.W., Pruessner, G., Chapman, S.C., Crosby, N.B.: 25 years of SOC: concepts and controversies. Space Sci. Rev. **198**(1), 3–44 (2016)
Weinberg, S.: Dreams of a Final Theory: The Scientist's Search for the Ultimate Laws of Nature. Vintage (1993)
Yano, I., Liu, C., Moncrieff, M.W.: Self-organized criticality and homeostasis in atmospheric convective organization. J. Atmos. Sci. **69**(12), 3449–3462 (2012)
Zeng, X., Pielke, R.A., Eykholt, R.: Chaos theory and its applications to the atmosphere. Bull. Am. Meteorol. Soc. **74**, 631–644 (1993)
Zhou, D., Gozolchiani, A., Ashkenazy, Y., Havlin, S.: Teleconnection paths via climate network direct link detection. Phys. Rev. Lett. **115**, 268501 (5 pages) (2015)

Chapter 4
Universal Inverse Power-Law Distribution for Temperature and Rainfall in the UK Region

Abstract Meteorological parameters, such as temperature, rainfall, pressure, etc., exhibit self-similar space-time fractal fluctuations generic to dynamical systems in nature such as fluid flows, spread of forest fires, earthquakes, etc. The power spectra of fractal fluctuations display inverse power-law form signifying long-range correlations. A general systems theory model predicts universal inverse power-law form incorporating the golden ratio for the fractal fluctuations. The model predicted distribution was compared with observed distribution of fractal fluctuations of all size scales (small, large and extreme values) in the historic monthwise temperature (maximum and minimum) and total rainfall for the four stations Oxford, Armagh, Durham and Stornoway in the UK region, for data periods ranging from 92 to 160 years. For each parameter, the two cumulative probability distributions, namely *c*max and *c*min starting from respectively maximum and minimum data value were used. The results of the study show that (i) temperature distributions (maximum and minimum) follow model predicted distribution except for Stornowy, minimum temperature *c*min. (ii) Rainfall distribution for *c*min follow model predicted distribution for all the four stations. (iii) Rainfall distribution for *c*max follows model predicted distribution for the two stations Armagh and Stornoway. The present study suggests that fractal fluctuations result from the superimposition of eddy continuum fluctuations.

Keywords Fractal fluctuations · Universal inverse power-law · UK region temperature and rainfall

4.1 Introduction

Dynamical systems such as fluid flows, heart beat patterns, spread of infectious diseases, etc., exhibit self-similar, i.e. a zigzag pattern of successive increases followed by decreases of all scales identified as fractal fluctuations. Self-similar structures incorporate in their geometrical design the Fibonacci numbers and thereby the golden mean or golden ratio τ ($\approx$1.618), exhibited vividly in plant

A.M. Selvam, *Self-organized Criticality and Predictability in Atmospheric Flows*, Springer Atmospheric Sciences, DOI 10.1007/978-3-319-54546-2_4

phyllotaxis, an intensive field of study in Botany (Jean 2009). The golden mean which underlies the Fibonacci series represents a nested continuum of self-similar structures or fractals implying long-range correlations or persistence between the larger and smaller scale structures. Self-similar shapes are everywhere and give the impression that they are being 'designed' to match the golden ratio. Bejan (2009) and Bejan and Peder (2012) show that these shapes emerge as part of an evolutionary phenomenon that facilitates the flow of information, in accordance with the constructal law of generation and evolution of design in nature. Fractal fluctuations signify non-local connections, i.e. long-range correlations in space and time manifested as inverse power-law form $f^{-\alpha}$ where f is the frequency (or wavelength of the eddies) and α the exponent. It is well known that power-law approximations fit a wide variety of physical and biological processes (Bejan and Peder 2012). Extensive studies by Lovejoy and Schertzer (2012), Bunde et al. (2013) have identified conclusively the self-similar fractal nature of fluctuations in meteorological parameters. The Gaussian probability distribution used widely for analysis and description of large data sets underestimates the probabilities of occurrence of extreme events such as stock market crashes, earthquakes, heavy rainfall, etc. The assumptions underlying the normal distribution such as fixed mean and standard deviation, independence of data, are not valid for real-world fractal data sets exhibiting a scale-free power-law distribution with fat tails (Selvam 2009). The author has developed a general systems theory model (Selvam 1990, 2007, 2009, 2011, 2014, 2015) for fractal fluctuations in dynamical systems. The model predicts universal inverse power-law form incorporating the golden mean for the probability distribution of amplitudes as well as variance (power spectra) of fractal fluctuations. The model predictions are compared with historic monthwise temperature (maximum and minimum) and total rainfall for the four stations Oxford, Armagh, Durham and Stornoway in the UK region for periods ranging from 160 to 92 years. A review of 1/*f* noise in physical phenomena, general systems theory model predictions for fractal fluctuations in dynamical systems are given in Chaps. 1 and 2. Details of data, analysis techniques and results are presented in the following.

4.2 Climate Variability

A characteristic feature of climatic records is their pronounced variability. The spectral analysis of continuous climatic time series normally reveals a continuous variance distribution encompassing all resolvable frequencies, with higher variance levels at lower frequencies or 'red noise' (Hasselmann 1976) typical of 1/*f* noise. Understanding the natural variability of climate is one of the most important tasks facing climatologists (Pelletier 1997).

Fraedrich (2002) and Fraedrich et al. (2009) give a brief summary of studies relating to 1/*f* noise in climate data. In the mid-1970s the Brownian motion entered climate research as an analog for the earth's climate fluctuations (Hasselmann 1976) which led to an intensive red noise search in data. At the same time observations and modeling of flicker noise and other power-law scaling regimes emerged (for example Voss and Clarke 1976; van Vliet et al. 1980) with concepts as close to the climate systems energy balance as the Brownian motion analog. Since then power-law power spectra different from Brownian motion has been identified in observed records and model simulations of the climate system. While most of these studies (for a review see Pelletier and Turcotte 1999) are guided by self-affine scaling laws governing the dynamics of a nonlinear system, the associated long-range memory or correlation aspect has been emphasized only recently (Koscielny-Bunde et al. (1998); Talkner and Weber (2000) analysing observed temperatures). The analyses suggest that the near surface temperature fluctuations are governed by a universal scaling behavior showing long-term memory correlations up to at least 30 years. Huybers and Curry (2006) demonstrate that climate variability exists at all timescales with climate processes being intimately coupled.

4.3 General Systems Theory for Fractal Fluctuations

Power (variance) spectra of fractal fluctuations exhibit inverse power-law form indicating (i) self-similar fractal fluctuations result from the coexistence of a continuum of eddies (waves) (ii) fractal fluctuations exhibit long-range space-time correlations since the amplitudes of larger and smaller size eddies are related to each other by the scale factor alone independent of other characteristics of the eddies.

The general systems theory model (Selvam 1990, 2007, 2009, 2011, 2014, 2015) is based on the above observational fact that fractal fluctuations signify an underlying eddy continuum. The model is based on the simple concept that large eddies result from successive space-time integration of enclosed small-scale fluctuations (eddies) analogous to Townsend's (1956) concept that large eddies are envelopes enclosing smaller scale eddies. The model predictions are as follows.

1. Fractal fluctuations result from the superimposition of eddy fluctuations of eddy continuum.
2. The eddy continuum growth occurs in two stages: (i) generation of primary dominant turbulent eddy of radius r (ii) large eddies form as envelopes enclosing these dominant turbulent eddies starting from unit primary eddy as zero level with length scale ratio z (R/r) equal to 1, 2, 3, etc. for successive stages of eddy growth. The primary eddy growth region is $z = 0$ to ± 1.
3. The eddy continuum traces an overall logarithmic spiral trajectory. The relationship between the circulation speeds W and w, respectively, of large and inherent small eddies of respective radii R and r is given by the logarithmic relationship

$$W = \frac{w}{k} \ln z \tag{4.1}$$

In (4.1) k is the steady state fractional volume dilution of the large eddy by internal smaller scale eddy mixing (Selvam 2013, 2014) and is given as

$$k = \sqrt{\frac{\pi}{2z}} \tag{4.2}$$

4. By concept W is the integrated mean of enclosed small-scale circulation speeds w. Therefore at each level, W represents the mean value associated with standard deviation equal to w. The normalized deviation t equal to mean/standard deviation is now given by W/w which is proportional to $\ln z$ from the above logarithmic relationship (Eq. 4.1). For fixed small eddy radius r, the normalized deviation t represents logarithm of eddy wavelengths (frequencies). The probability P distributions of amplitudes and/or variance of eddy fluctuations when plotted with normalized deviation t will represent the conventional plot with logarithm of eddy wavelength (frequency) on x-axis versus probability P on y-axis. Such a plot of probability distributions P of amplitude and also variance of fractal fluctuations displays inverse power-law form discussed at items five and six in the following.
5. The probability distribution P is a function of the golden mean τ. The golden mean which underlies the Fibonacci series represents a nested continuum of self-similar structures or fractals implying long-range correlations or persistence between the larger and smaller scale structures and is consistent with observed inverse power-law distribution for P. For the range of normalized deviation t values $t \geq 1$ and $t \leq -1$, P is given as

$$P = \tau^{-4t} \tag{4.3}$$

6. The primary eddy growth corresponds to normalized deviation t ranging from -1 to $+1$. In this region the probability P is a function of the steady state fractional volume dilution k of the growing primary eddy by internal smaller scale eddy mixing (Selvam 2013, 2014) (Eq. 4.2). The expressions for P is given as

$$P = \tau^{-4k} \tag{4.4}$$

7. The model predicted universal inverse power-law distribution is very close to statistical normal distribution for normalized deviation $t < 2$ and $t > -2$ and exhibits a long fat tail for $t \geq 2$ and $t \leq -2$, i.e. extreme events have a higher probability of occurrence than that predicted by statistical normal distribution as

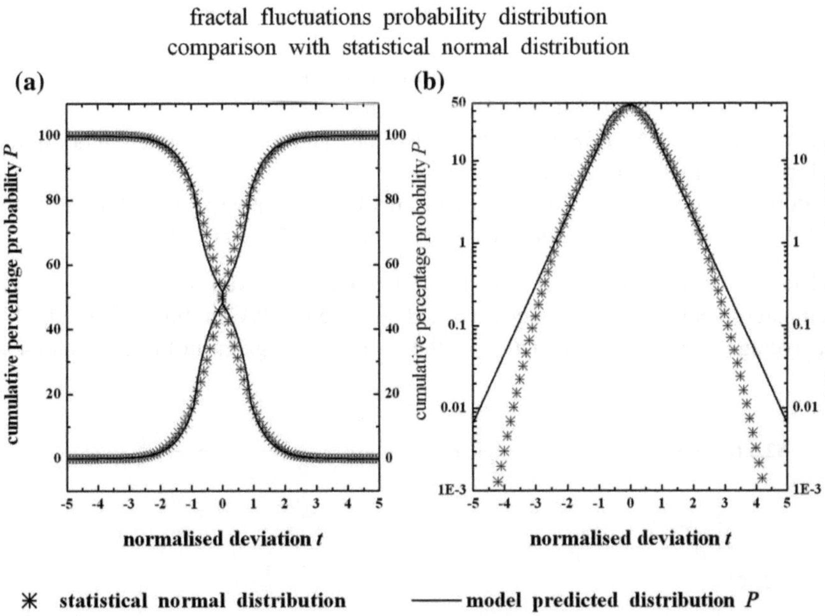

Fig. 4.1 Model predicted and statistical normal cumulative probability distribution P plotted with respect to normalized deviation t. Panel A shows on linear scale t on x-axis and P on y-axis. In Panel B, the logarithmic scale for P is plotted up to 50% only and shows on a magnified scale the small values (<2.5%) of P for $t \geq 2$ and $t \leq -2$ in the region of extreme events where there is an appreciable departure between model predicted and statistical normal distribution

found in practice. The statistical normal distribution and the model predicted universal inverse power-law distribution are shown in Fig. 4.1 (Selvam 2013, 2014).

8. The general systems theory model shows that (i) the probability distribution P as given in Eqs. 4.3 and 4.4 when plotted with respect to normalized deviation t gives unique and universal quantification for the observed inverse power-law distribution for power (variance) spectrum of fractal fluctuations (ii) the amplitude distribution of fractal fluctuations is also quantified by the same probability distribution P (Selvam 1990, 2007, 2014, 2015).

4.4 Data

Monthwise climate data of temperature (maximum and minimum) and rainfall for four stations (i) Oxford (ii) Armagh (iii) Durham and (iv) Stornoway in the UK region were obtained from http://www.metoffice.gov.uk/climate/uk/stationdata.

The monthwise climate data used are as follows:

- Oxford (51.75°N, 1.26°W, 63 m amsl) max and min temperature and rainfall for 160 years (1853–2012)
- Armagh (54.35°N, 6.65°W, 62 m amsl), max and min temperature 147 years (1866–2012) and rainfall for 160 years (1853–2012)
- Durham (54.78°N, 1.58°W, 102 m amsl) max and min temperature and rainfall for 133 years (1880–2012)
- Stornoway (58.21°N, 6.39°W, 15 m amsl) max temperature and rainfall for 139 years (1874–2012) and min temperature for 92 years (1921–2012)

The regional locations of the four stations are given in Fig. 4.2 (http://ukclimateprojections.metoffice.gov.uk/media/image/6/q/WGen_FigA2_large.jpg). The average and standard deviation of the data sets are given in Figs. 4.3 and 4.4.

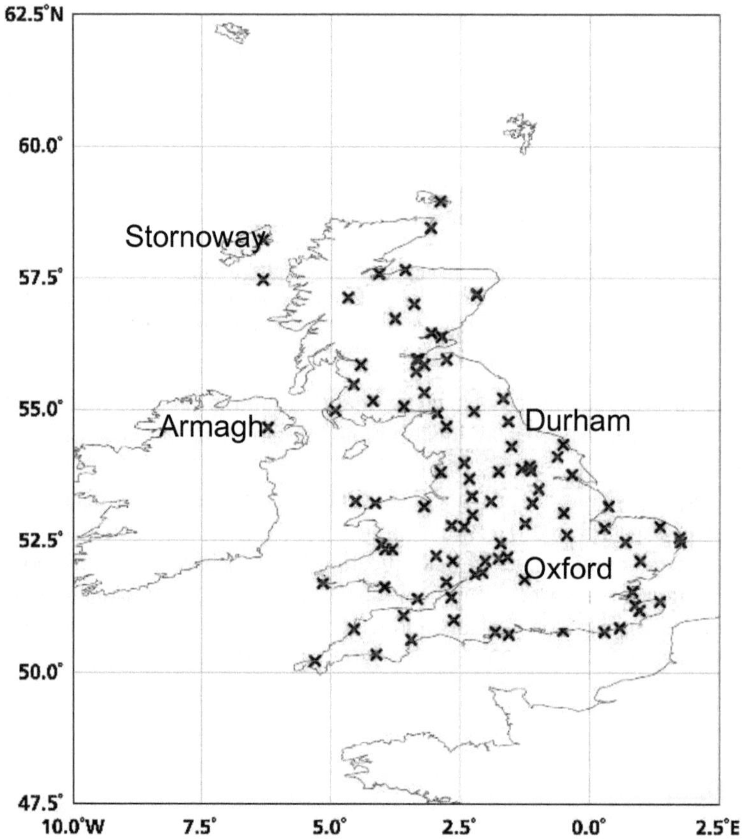

Fig. 4.2 Location of Oxford, Armagh, Durham and Stornoway in the UK region (http://ukclimateprojections.metoffice.gov.uk/media/image/6/q/WGen_FigA2_large.jpg)

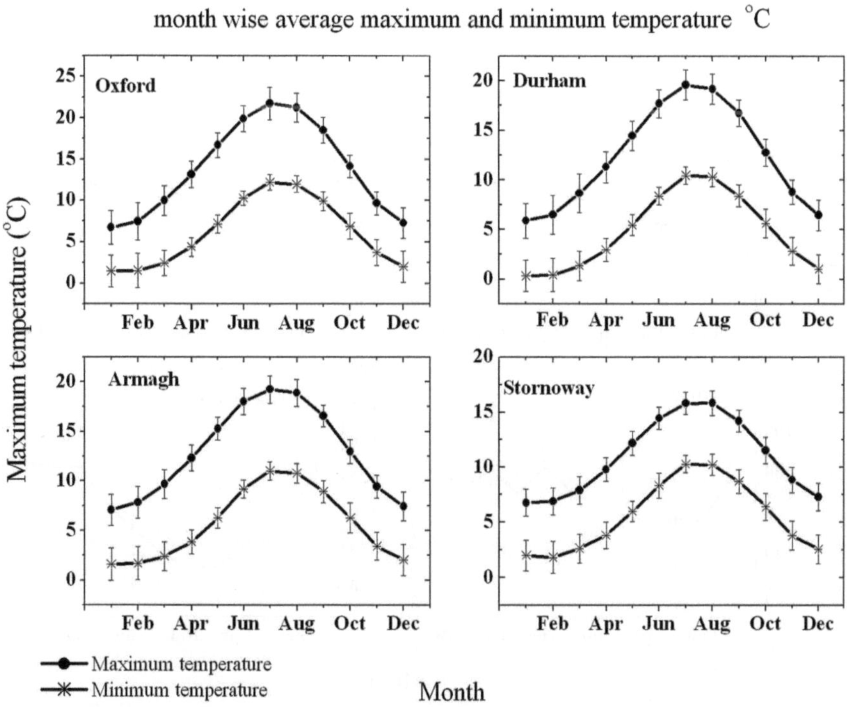

Fig. 4.3 Monthwise average and standard deviation of maximum and minimum temperatures for Oxford, Armagh, Durham and Stornoway in the UK region

4.5 Analysis

Each data set was represented as the frequency of occurrence *f*(*i*) in a suitable number *n* of class intervals *x*(*i*), *i* = 1, *n* covering the range of values from *minimum* to the *maximum* in the data set. The class interval *x*(*i*) represents dataset values in the range $x(i) \pm \Delta x$, where Δx is a constant. The average av and standard deviation sd for the data set is computed as

$$\text{av} = \frac{\sum_1^n [x(i) \times f(i)]}{\sum_1^n f(i)} \tag{4.5}$$

$$\text{var} = \frac{\sum_1^n \left\{[x(i) - \text{av}]^2 \times f(i)\right\}}{\sum_1^n f(i)} \tag{4.6}$$

$$\text{sd} = \text{sqrt}(\text{var})$$

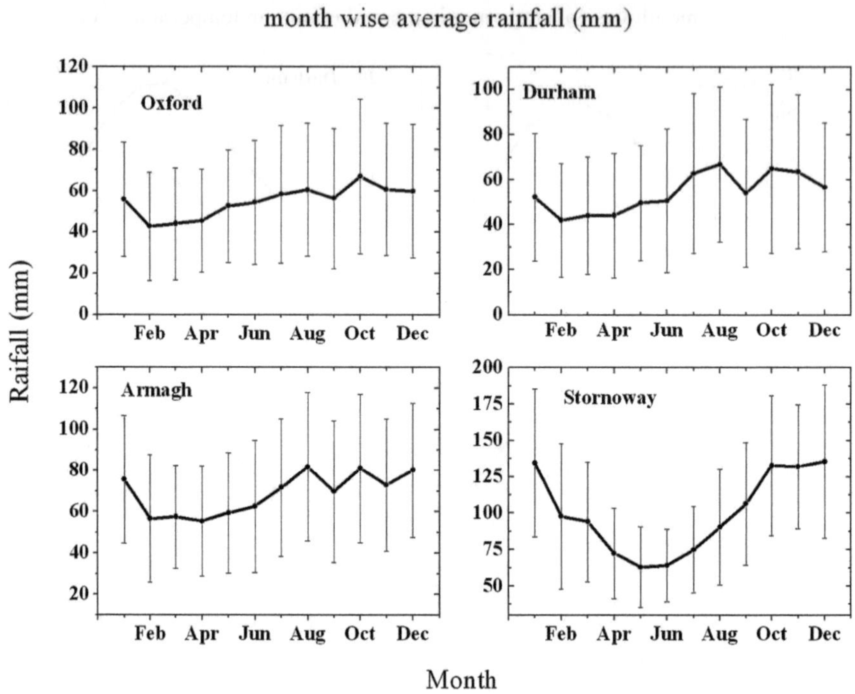

Fig. 4.4 Monthwise average and standard deviation of rainfall for Oxford, Armagh, Durham and Stornoway in the UK region

The *normalized deviation t* values for class intervals *t*(*i*) were then computed as

$$t(i) = \frac{x(i) - \mathrm{av}}{\mathrm{sd}} \tag{4.7}$$

The cumulative percentage probabilities of occurrence *c*max(*i*) and *c*min(*i*) were then computed starting respectively from the maximum ($i = n$) and minimum ($i = 1$) class interval values as follows:

$$c\max(i) = \frac{\sum_n^i [x(i) \times f(i)]}{\sum_1^n [x(i) \times f(i)]} \times 100.0 \tag{4.8}$$

$$c\min(i) = \frac{\sum_1^i [x(i) \times f(i)]}{\sum_1^n [x(i) \times f(i)]} \times 100.0. \tag{4.9}$$

The monthwise (January to December) average and standard deviation of cumulative percentage probability values *c*max(*i*) and *c*min(*i*) for temperature and rainfall were computed for the four UK stations and plotted with respect to

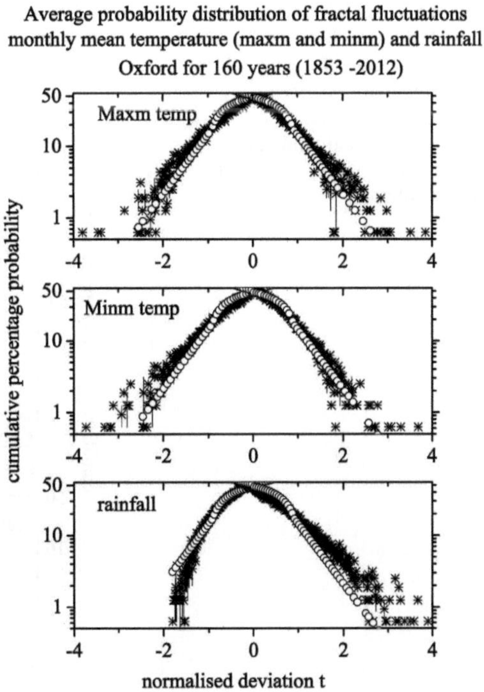

Fig. 4.5 Average cumulative probability distributions (*) *c*min and *c*max of fractal fluctuations for Oxford. Temperature distributions (both *c*min and *c*max) follow model predicted (O) distribution. Rainfall distribution is significant only for *c*min

corresponding *normalized deviation* $t(i)$ values with logarithmic scale for the probability axis (Figs. 4.5, 4.6, 4.7, 4.8) along with model predicted universal inverse power-law distribution. The results are presented in Sect. 4.6

4.6 Results and Discussion

The probability distribution P of amplitudes of fractal fluctuations in monthwise (January–December) temperature (maximum and minimum) and total rainfall for fluctuations of all size scales (small, large and extreme) was compared with the general systems theory model predicted universal inverse power-law distribution $P = \tau^{-4t}$. Chi-square test was applied to test the 'Goodness of fit' at 5% significance level for the *c*min and *c*max distributions. The following results were obtained.

i. Temperature distributions (*c*min and *c*max) follow model predicted distribution except for Stornowy, minimum temperature *c*min.
ii. Rainfall distribution *c*min follows model predicted distribution for all the four stations.
iii. Rainfall distribution *c*max follows model predicted distribution for the two stations Armagh and Stornoway.

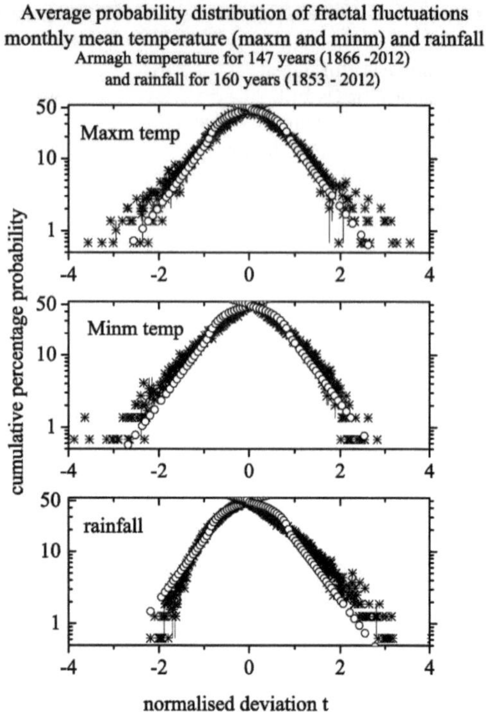

Fig. 4.6 Average cumulative probability distributions (*) *c*min and *c*max of fractal fluctuations for Armagh. Temperature and rainfall distributions (both *c*min and *c*max) follow model predicted (O) distribution

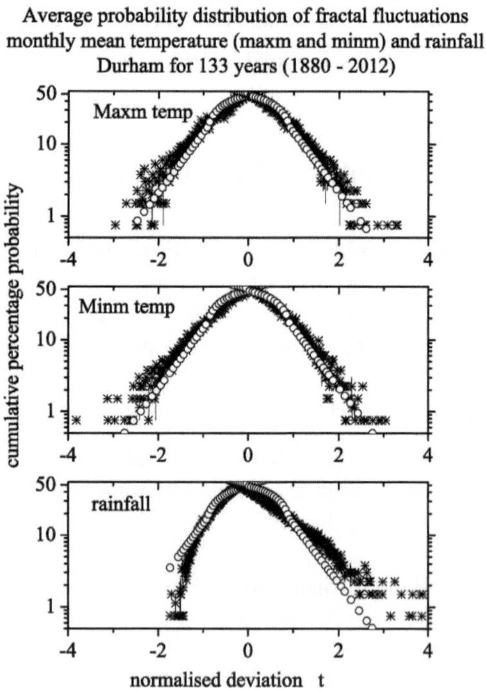

Fig. 4.7 Average cumulative probability distributions (*) *c*min and *c*max of fractal fluctuations for Durham. Temperature distributions (both *c*min and *c*max) follow model predicted (O) distribution. Rainfall distribution is significant only for *c*min

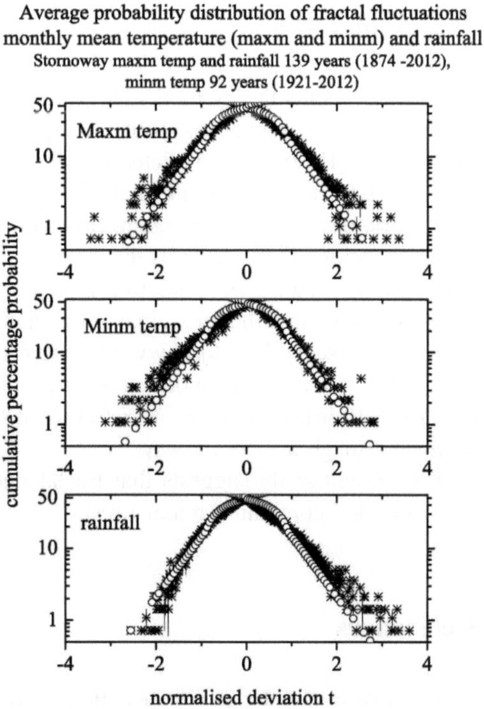

Fig. 4.8 Average cumulative probability distributions (*) *c*min and *c*max of fractal fluctuations for Stornoway. Maximum temperature and rainfall distributions (both *c*min and *c*max) follow model predicted (O) distribution. Minimum temperature distribution is significant only for *c*max

Extreme events correspond to normalized deviation $t \geq 2$ and $t \leq -2$ and in this region the number of data points with mean and standard deviation for probability of occurrence at each t value, i.e. at least two samples, is small for temperature and rainfall. The observed mean probability of occurrence was considered to agree with model predicted (5% significance level) if the model predicted lies within two standard deviations from mean of the observed. For temperature, more than 50% of the extreme values agree with model prediction. Only positive extreme values are available for rainfall and 20 to 45% agree with model prediction.

Inverse power-law distribution for fractal fluctuations implies long-range space-time correlations manifested as memory or persistence in the space-time variability of the meteorological parameter such as rainfall, temperature, etc. Kantelhardt et al. (2006) stated that the persistence analysis of river flows and precipitation has been initiated, about half a century ago, by H.E. Hurst, who found that runoff records from various rivers exhibit 'long-range statistical dependencies' (Hurst 1951). Later, similar long-term correlated fluctuation behavior has also been reported for many other geophysical records including temperature and precipitation data (Kantelhardt et al. 2006). Characterizing and understanding the persistence of wet and dry conditions in the distant past gives new perspectives on contemporary climate change and its causes (Bunde et al. 2013).

4.7 Conclusion

A general systems theory model (Selvam 1990, 2007, 2009, 2014, 2015) predicts universal inverse power-law form incorporating the golden mean for the fractal fluctuations. The model predicted distribution was compared with observed distribution of fractal fluctuations of all size scales (small, large and extreme values) in the historic monthwise temperature (maximum and minimum) and total rainfall for the four stations Oxford, Armagh, Durham and Stornoway in the UK region, for data periods ranging from 92 to 160 years. The results of the study show that (i) temperature distributions (maximum and minimum) follow model predicted distribution except for Stornowy, minimum temperature *c*min. (ii) Rainfall distribution for *c*min follow model predicted distribution for all the four stations. (iii) Rainfall distribution for *c*max follow model predicted distribution for the two stations Armagh and Stornoway.

The present study suggests that fractal fluctuations result from the superimposition of eddy continuum fluctuations.

References

Bejan, A., Peder, Z.J.: Design in nature. How the constructal law governs evolution in biology, physics, technology, and social organization. Double Day, London, UK, (2012)

Bejan, A.: The golden ratio predicted: vision, cognition and locomotion as a single design in nature. Int. J. Design and Nature and Ecodyn **4**(2), 97–104 (2009)

Bunde, A., Büntgen, U., Ludescher, J., Luterbacher, J., von Storch, H.: Is there memory in precipitation? Nat. Clim. Change **3**, 174–175 (2013)

Fraedrich, K.: Fickian diffusion and Newtonian cooling: a concept for noise induced climate variability with long-term memory? Stochast. Dyn. **2**, 403–412 (2002)

Fraedrich, K., Blender, R., Zhu, X.: Continuum climate variability: long-term memory, scaling, and (1/*f*)-noise. Int. J. Mod. Phy. B **23**(28 and 29), 5403–5416 (2009)

Hasselmann, K.: Stochastic climate models: part I. Theory. Tellus. **28**(6), 473–485 (1976)

Hurst, H.E.: Long-term storage capacity of reservoirs. Trans. Am. Soc. Civ. Eng. **116**, 770–799 (1951)

Huybers, P., Curry, W.: Links between annual, Milankovitch and continuum temperature variability. Nature **441**, 329–332 (2006)

Jean, R.V.: Phyllotaxis: a systemic study in plant morphogenesis. Cambridge University Press, Cambridge (2009)

Kantelhardt, J.W., Koscielny-Bunde, E., Rybski, D., Braun, P., Bunde, A., Havlin, S.: Long-term persistence and multifractality of precipitation and river runoff records. J. Geophy. Res. **111**, D01106 (2006)

Koscielny-Bunde, E., Bunde, A., Havlin, S., Roman, H.E., Goldreich, Y., Schellnhuber, H.-J.: Indication of a universal persistence law governing atmospheric variability. Phys. Rev. Lett. **81** (3), 729–732 (1998)

Lovejoy, S., Schertzer, D.: The weather and climate: emergent laws and multifractal cascades. Cambridge University Press, Cambridge (2012)

Pelletier, J.D.: Analysis and modeling of the natural variability of climate. J. Clim. **10**, 1331–1342 (1997)

Pelletier, J.D., Turcotte, D.L.: Self-affine time series: II. Applications and models. advances in geophysics **40**, 91–166 (1999)

Selvam A.M.: Deterministic chaos, fractals and quantumlike mechanics in atmospheric flows. Can. J. Phys. **68:**831–841 (1990). http://xxx.lanl.gov/html/physics/0010046

Selvam, A.M.: Chaotic Climate Dynamics. Luniver Press, UK (2007)

Selvam, A.M.: Fractal fluctuations and statistical normal distribution. Fractals **17**(3):333–349 (2009). http://arxiv.org/pdf/0805.3426

Selvam, A.M.: Signatures of universal characteristics of fractal fluctuations in global mean monthly temperature anomalies. J. Syst. Sci. Complexity **24**(1), 14–38 (2011)

Selvam, A.M.: Scale-free universal spectrum for atmospheric aerosol size distribution for Davos, Mauna Loa and Izana. Int. J. Bifurcat Chaos **23**(2), 1350028 (13 pages) (2013)

Selvam, A.M.: Universal inverse power-law distribution for temperature and rainfall in the UK region. Dyn. Atmos. Oceans **66**, 138–150 (2014)

Selvam, A.M.: Rain formation in warm clouds: general systems theory. Springer Briefs in Meteorology, Springer (2015)

Talkner, P., Weber, R.O.: Power spectrum and detrended fluctuation analysis: application to daily temperatures. Phys. Rev. E **62**(1), 150–160 (2000)

Townsend, A.A.: The structure of turbulent shear flow, 2nd edn, pp. 115–130. Cambridge University Press, London, UK (1956)

van Vliet, K.M., van der Ziel, A., Schmidt, R.R: Temperature-fluctuation noise of thin films supported by a substrate. J. Appl. Phys. **51**, 2947 (1980) (10 pages)

Voss, R.F., Clarke, J.: Flicker (1/*f*) noise: Equilibrium temperature and resistance fluctuations. Phys. Rev. B **13**, 556–573 (1976)

Chapter 5
Signatures of Universal Characteristics of Fractal Fluctuations in Global Mean Monthly Temperature Anomalies

Abstract Self-similar space-time fractal fluctuations are generic to dynamical systems in nature such as atmospheric flows, heartbeat patterns, population dynamics, etc. The physics of the long-range correlations intrinsic to fractal fluctuations is not completely understood. It is important to quantify the physics underlying the irregular fractal fluctuations for prediction of space-time evolution of dynamical systems. A general systems theory model for fractals visualising the emergence of successively larger-scale fluctuations resulting from the space-time integration of enclosed smaller scale fluctuations predicts the following. (i) The probability distribution and the power spectrum for fractal fluctuations is the same inverse power-law function incorporating the *golden mean*. (ii) The predicted distribution is close to the Gaussian distribution for small-scale fluctuations but exhibits *fat long tail* for large-scale fluctuations with higher probability of occurrence than predicted by Gaussian distribution. (iii) Since the power spectrum (variance, i.e. square of eddy amplitude) also represents the probability densities as in the case of quantum systems such as the electron or photon, fractal fluctuations exhibit quantum-like chaos. (iv) The *fine-structure constant* for spectrum of fractal fluctuations is a function of the golden mean and is analogous to atomic spectra equal to about 1/137. Global gridded time series data sets of monthly mean temperatures for the period 1880—2007/2008 were analysed. The data sets and the corresponding power spectra exhibit distributions close to the model predicted inverse power-law distribution. The model predicted and observed universal spectrum for interannual variability rules out linear secular trends in global monthly mean temperatures. Global warming results in intensification of fluctuations of all scales and manifested immediately in high frequency fluctuations.

Keywords Fractals and statistical normal distribution · Power-law distributions · Long-range correlations and fat tail distributions · Golden mean and fractal fluctuations

A.M. Selvam, *Self-organized Criticality and Predictability in Atmospheric Flows*, Springer Atmospheric Sciences, DOI 10.1007/978-3-319-54546-2_5

5.1 Introduction

Dynamical systems in nature such as atmospheric flows, heartbeat patterns, population dynamics, stock market indices, DNA base A, C, G, T sequence pattern, etc., exhibit irregular space-time fluctuations on all scales and exact quantification of the fluctuation pattern for predictability purposes has not yet been achieved. Traditionally, the Gaussian probability distribution is used for a broad quantification of the data set variability in terms of the sample mean and variance. The fractal or self-similar nature of space-time fluctuations was identified by Mandelbrot (1975) in the 1970s. Fractal fluctuations show a zigzag self-similar pattern of successive increase followed by decrease on all scales (space-time), for example, in atmospheric flows, cycles of increase, and decrease in meteorological parameters such as windspeed, temperature, etc., occur from the turbulence scale of millimeters-seconds to climate scales of thousands of kilometers-years. The power spectra of fractal fluctuations exhibit inverse power law of the form $f^{-\alpha}$ where f is the frequency and α is a constant. Inverse power law for power spectra indicates long-range space-time correlations or scale invariance for the scale range for which α is a constant, i.e. the amplitudes of the eddy fluctuations in this scale range are a function of the scale factor α alone. In general, the value of α is different for different scale ranges indicating multifractal structure for the fluctuations. The long-range space-time correlations exhibited by dynamical systems are identified as self-organized criticality (Bak et al. 1988; Schroeder 1991). The physics of self-organized criticality is not yet identified. The physics of fractal fluctuations generic to dynamical systems in nature is not yet identified and traditional statistical, mathematical theories do not provide adequate tools for identification and quantitative description of the observed universal properties of fractal structures observed in all fields of science and other areas of human interest. A general systems theory for fractal space-time fluctuations (Selvam 1990, 2007, 2015; Selvam and Fadnavis 1998) shows that the larger-scale fluctuation can be visualized to emerge from the space-time averaging of enclosed small-scale fluctuations, thereby generating a hierarchy of self-similar fluctuations manifested as the observed eddy continuum in power spectral analyses of fractal fluctuations. Such a concept results in inverse power law form incorporating the golden mean τ for the space-time fluctuation pattern and also for the power spectra of the fluctuations. The predicted distribution is close to the Gaussian distribution for small-scale fluctuations, but exhibits *fat long tail* for large-scale fluctuations. Analyses of extensive data sets of Global gridded data sets of monthly mean temperatures for the period 1880—2007/2008 show that the space/time data sets follow closely, but not exactly the statistical normal distribution, particularly in the region of normalized deviations t greater than 2, the t values being computed as equal to $(x - av)/sd$ where av and sd denote respectively the mean and standard deviation of the variable x. The general systems theory, originally developed for turbulent fluid flows, provides universal quantification of physics underlying fractal fluctuations and is applicable to all dynamical systems in nature independent of its physical, chemical, electrical, or any other intrinsic characteristic. A review of *1/f* noise in physical phenomena, general systems theory model predictions for fractal fluctuations in dynamical systems are given in Chaps. 1 and 2. Details of data, analysis techniques and results are presented in the following.

5.2 Scale-Free Theory for Power Laws with Fat, Long Tails

The general systems theory, a scale-free theory for the observed fractal fluctuations in atmospheric flows, show that the observed long-range spatiotemporal correlations are intrinsic to quantum-like chaos governing fluid flows. The model concepts are independent of the exact details such as the chemical, physical, physiological, and other properties of the dynamical system and therefore provide a general systems theory applicable to all real world and computed dynamical systems in nature (Selvam 1990, 2001a, b, 2002a, b, 2004, 2007, 2015; Selvam et al. 2000; Selvam and Fadnavis 1998, 1999). The model is based on the concept that the irregular fractal fluctuations may be visualized to result from the superimposition of an eddy continuum, i.e. a hierarchy of eddy circulations generated at each level by the space-time integration of enclosed small-scale eddy fluctuations. Such a concept of space-time fluctuation averaged distributions *should* follow statistical normal distribution according to *Central Limit Theorem* in traditional Statistical theory (Ruhla 1992). Also, traditional statistical/mathematical theory predicts that the Gaussian, its Fourier transform and therefore Fourier transform associated power spectrum are the same distributions. The Fourier transform of normal distribution is essentially a normal distribution. A power spectrum is based on the Fourier transform, which expresses the relationship between time (space) domain and frequency domain description of any physical process (Phillips 2005; Riley et al. 2006). However, the general systems theory model visualizes the eddy growth process in successive stages of unit length step growth with ordered two-way energy feedback between the larger and smaller scale eddies and derives a power-law probability distribution P which is close to the Gaussian for small deviations and gives the observed fat, long tail for large fluctuations. Further, the model predicts the power (variance) spectrum of the eddy continuum also to follow the power-law probability distribution P.

In summary, the model predicts the following:

- The eddy continuum consists of an overall logarithmic spiral trajectory with the quasiperiodic *Penrose* tiling pattern for the internal structure.
- The successively larger eddy space-time scales follow the Fibonacci number series.

- The probability distribution P of fractal domains for the nth step of eddy growth is equal to τ^{-4n} where τ is the golden mean equal to $(1 + \sqrt{5})/2$ ($\approx$1.618). The eddy growth step n represents the *normalized deviation t* in traditional statistical theory. The *normalized deviation t* represents the departure of the variable from the mean in terms of the standard deviation of the distribution assumed to follow *normal distribution* characteristics for many real-world space-time events. There is progressive decrease in the probability of occurrence of events with increase in corresponding *normalized deviation t*. Space-time events with *normalized deviation t* greater than 2 occur with a probability of 5% or less and may be categorized as extreme events associated in general with widespread (space-time) damage and loss. The model predicted probability distribution P is close to the statistical normal distribution for t values less than 2 and greater than normal distribution for t more than 2, thereby giving a *fat, long tail*. There is nonzero probability of occurrence of very large events.
- The inverse of probability distribution P, namely, τ^{4n} represents the relative eddy energy flux in the large eddy fractal (small-scale fine structure) domain. There is progressive decrease in the probability of occurrence of successive stages of eddy growth associated with progressively larger domains of fractal (small-scale fine structure) eddy energy flux and at sufficiently large growth stage trigger catastrophic extreme events such as heavy rainfall, stock market crashes, traffic jams, etc., in real-world situations.
- The power (variance) spectrum of fractal fluctuations also follows the same distribution P as for the distribution of fractal fluctuations. The square of the eddy amplitude (variance) represents the eddy energy and therefore the eddy probability density P. Such a result that the additive amplitudes of eddies when squared represent probabilities, is exhibited by the subatomic dynamics of quantum systems such as the electron or proton (Maddox 1988, 1993; Rae 1988). Therefore fractal fluctuations are signatures of quantum-like chaos in dynamical systems. The above-described analogy of quantum-like mechanics for dynamical systems is similar to the concept of a subquantum level of fluctuations whose space-time organization gives rise to the observed manifestation of subatomic phenomena, i.e. quantum systems as order out of chaos phenomena (Grossing 1989).
- The *fine structure constant* for spectrum of fractal fluctuations is a function of the *golden mean* and is analogous to that of atomic spectra equal to about 1/137.
- The universal algorithm for self-organized criticality is expressed in terms of the universal *Feigenbaum*'s constants (Feigenbaum 1980) $\boldsymbol{a}$ and $\boldsymbol{d}$ as $2a^2 = \pi d$ where the fractional volume intermittency of occurrence $\boldsymbol{\pi d}$ contributes to the total variance $2\boldsymbol{a}^2$ of fractal structures. The *Feigenbaum*'s constants are expressed as functions of the *golden mean*. The probability distribution P of fractal domains is also expressed in terms of the Feigenbaum's constants $\boldsymbol{a}$ and $\boldsymbol{d}$. The details of the model are summarized in Chap. 3.

5.3 Data and Analysis

5.3.1 Data

Gridded temperature anomalies for mean temperatures obtained from the GHCN V2 (http://www.ncdc.noaa.gov/oa/climate/ghcn-monthly/index.php and ftp.ncdc.noaa.gov/pub/data/ghcn/v2/) monthly temperature data sets for 129 years (1880–2008) the months January to May and for 128 years (1880–2008) for the months June to December were used for the study. Details of the data set as given in the 'README_GRID_TEMP' text document are as follows. GHCN homogeneity adjusted data was the primary source for developing the gridded fields. In grid boxes without homogeneity adjusted data, GHCN raw data was used to provide additional coverage when possible. Each month of data consists of 2592 gridded data points produced on a 5 × 5 degree basis for the entire globe (72 longitude × 36 latitude grid boxes).

Gridded data for every month from January 1880 to the most recent month (May 2008 in the present study) is available. The data are temperature anomalies in degrees Celsius. Each gridded value was multiplied by 100 and written to file as an integer. Missing values are represented by the value −32,768.

There are two options: grid_1880_YYYY.dat.gz (used in the present study), where YYYY is the current year and the values are calculated using the first difference method, and anom-grid-1880-current.dat.gz which uses the 'anomaly' method.

The data are formatted by year, month, latitude, and longitude. There are twelve longitude grid values per line, so there are 6 lines (72/12 = 6) for each of the 36 latitude bands. Longitude values are written from 180 W to 180 E, and latitude values from 90 N to 90 S. Data for each month is preceded by a label containing the month and year of the gridded data.

```
for year = begyr to endyr
  for month = 1 to 12
    format(2i5) month, year
    for ylat = 1 to 36 (85–90 N, 80–85 N, …, 80–85 S, 85–90 S)
      format(12i5) 180–175 W, 175–170 W, …, 130–125 W, 125–120 W
      format(12i5) 120–115 W, 175–170 W, …, 70–65 W, 65–60 W
      format(12i5) 60–55 W, 55–50 W, …, 10–5 W, 5–0 W
      format(12i5) 0–5 E, 5–10 E, …, 50–55 E, 55–60 E
      format(12i5) 60–65 E, 65–70 E, …, 110–115 E, 115–120 E
      format(12i5) 120–125 E, 125–130 E, …, 170–175 E, 175–180 E
```

Each file has been compressed using 'gzip'. They can be uncompressed with 'WinZip' for those using Windows 95 (and above) or with 'gzip' from most UNIX platforms. The FORTRAN utility program 'read_gridded.f' can be downloaded to assist in extracting data of interest. This program allows the user to extract nonmissing values for selected months and write the data to an ascii output file. The latitude and longitude of the center of each corresponding grid box accompanies each gridded value in the output file.

These anomalies were calculated with respect to the period 1961–1990 using the First Difference Method, an approach developed to maximize the use of available station records (Peterson and Vose 1997; Peterson et al. 1998a, b). The First Difference Method involves calculating a series of calendar-month differences in temperature between successive years of station data (FDyr = Tyr–Tyr−1). For example, when creating a station's first difference series for mean February temperature, we subtract the station's February 1880 temperature from the station's February 1881 temperature to create a February 1881 first difference value. First difference values for subsequent years are calculated in the same fashion by subtracting the station's preceding year temperature for all available years of station data.

For each year and month the sum of the 'first difference' value of all stations located within the appropriate 5 × 5 degree box was determined and divided by the total number of stations in the box to get an unweighted first difference value for each grid box. Next calculate a cumulative sum of these gridded first difference values for all years from 1880 to 1998 to produce a time series for each grid box. The cumulative sum is calculated for each grid box and each month of gridded first difference data independently through time. Each grid box time series is then adjusted to create anomalies with respect to the period 1961–1990.

This gridded data set was developed to produce the most accurate time series possible. However, this required to treat months and grid boxes independently through time. The use of this data is most appropriate for analysing the change in temperature within a particular grid box, or set of grid boxes, over a span of years. If one is more interested in analysing temperature changes within individual years, e.g. the change in temperature between February and March, 1908, or between two regions in 1908, it is recommended that the GHCN station data be used directly.

5.4 Analyses and Results

5.4.1 Frequency Distribution

Each data set (monthwise temperature time series for the period 1800 to 2008 for the months January to May and for the period 1800 to 2007 for the months June to December) was represented as the frequency of occurrence $f(i)$ in a suitable number n of class intervals $x(i)$, $i = 1, n$ covering the range of values from *minimum* to the *maximum* in the data set. The class interval $x(i)$ represents dataset values in the range $x(i) \pm \Delta x$, where Δx is a constant. The average *av* and standard deviation *sd* for the data set is computed as

$$av = \frac{\sum_1^n [x(i) \times f(i)]}{\sum_1^n f(i)}$$

$$sd = \frac{\sum_1^n \left\{ [x(i) - av]^2 \times f(i) \right\}}{\sum_1^n f(i)}$$

The *normalized deviation t* values for class intervals $t(i)$ were then computed as

$$t(i) = \frac{x(i) - av}{sd}.$$

The cumulative percentage probabilities of occurrence cmax(i) and cmin(i) corresponding to the *normalized deviation t* values were then computed starting, respectively, from the maximum ($i = n$) and minimum ($i = 1$) class interval values as follows:

$$\text{cmax}(i) = \frac{\sum_n^i [x(i) \times f(i)]}{\sum_1^n [x(i) \times f(i)]} \times 100.0$$

$$\text{cmin}(i) = \frac{\sum_n^i [x(i) \times f(i)]}{\sum_1^n [x(i) \times f(i)]} \times 100.0.$$

The cmax and cmin distributions with respect to the *normalized deviation t* values were computed for each month for all grid points including grid points which did not have continuous time series data. The total number of grid points available for the study for the months January to December is given in Fig. 5.1. The average and standard deviation of cmax(i) and cmin(i) for each *normalized deviation* $t(i)$ values were then computed for each month from the corresponding distributions for all the available grid points for the month. The average cumulative percentage probability values cmax(i) and cmin(i) plotted with respect to corresponding *normalized deviation* $t(i)$ values with linear axes are shown in Fig. 5.2 and on logarithmic scale for the probability axis in the tail region, i.e. *normalized deviation* $t(i)$ values greater than 2 in Fig. 5.3 for positive extremes $t(i) = 2$ to 4, and in Fig. 5.4 for negative extremes $t(i) = -2$ to -4. The standard deviation of each mean cmax(i) and cmin(i) value is shown as a vertical error bar on either side of the mean in Figs. 5.2, 5.3, and 5.4. The figures also contain the statistical normal distribution and the computed theoretical distribution (Sect. 5.2) for comparison. The figures show clearly the appreciable positive departure of observed probability densities from the statistical normal distribution for extreme values at normalized deviation t values more than 2 (Figs. 5.3 and 5.4). The observed extreme values corresponding to $t(i)$ values greater than 2 for cmax(i) and cmin(i) distributions were compared for 'goodness of fit' with computed theoretical distribution and statistical normal distribution as follows. For cmax(i) and cmin(i) values, where standard deviation is available (no of observed values more than one), if the

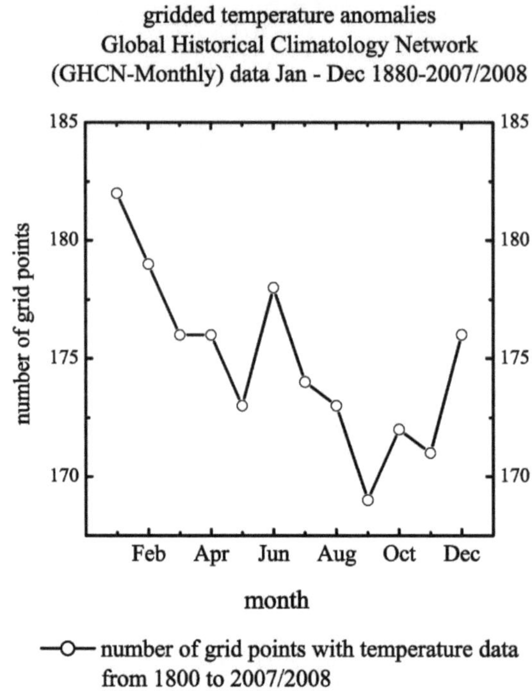

Fig. 5.1 The total number of grid points available for the study for the months January to December

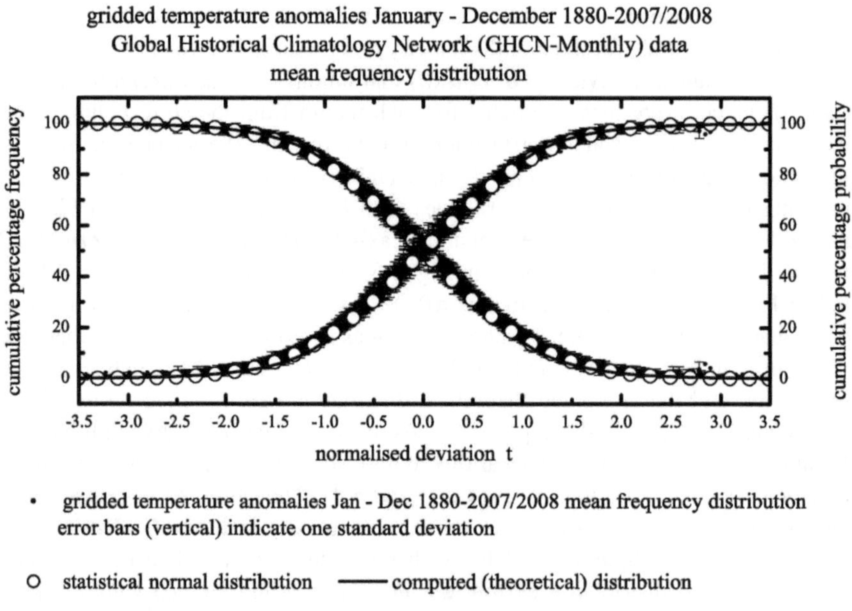

Fig. 5.2 Average cumulative percentage probability distribution

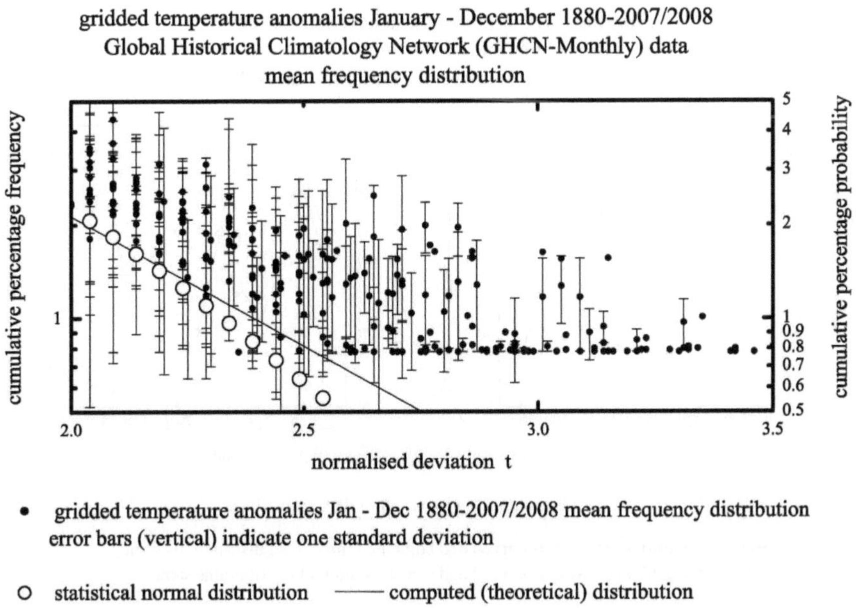

Fig. 5.3 Average cumulative percentage probability values on logarithmic scale for the probability axis in the positive tail region (extreme events), i.e. *normalized deviation t* values greater than 2

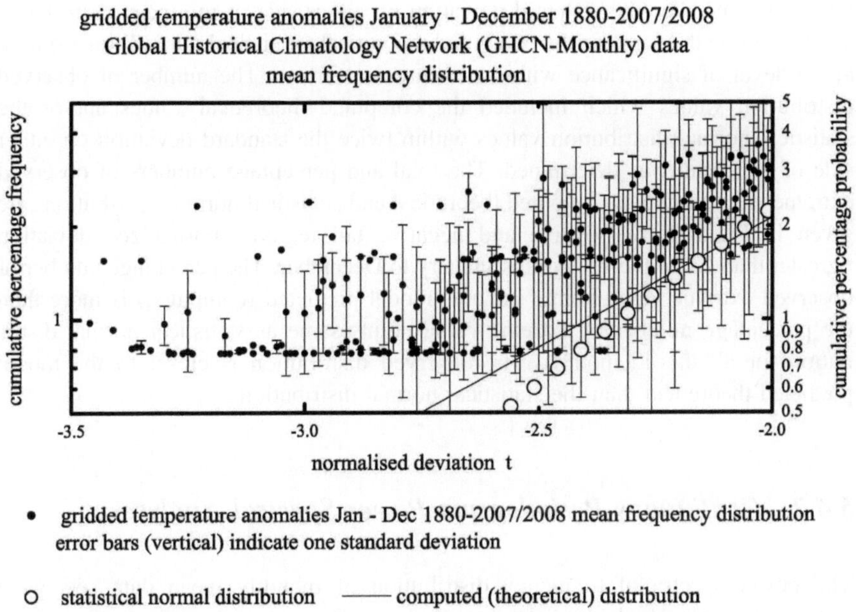

Fig. 5.4 Average cumulative percentage probability values on logarithmic scale for the probability axis in the negative tail region (extreme events), i.e. *normalized deviation t* values greater than −2

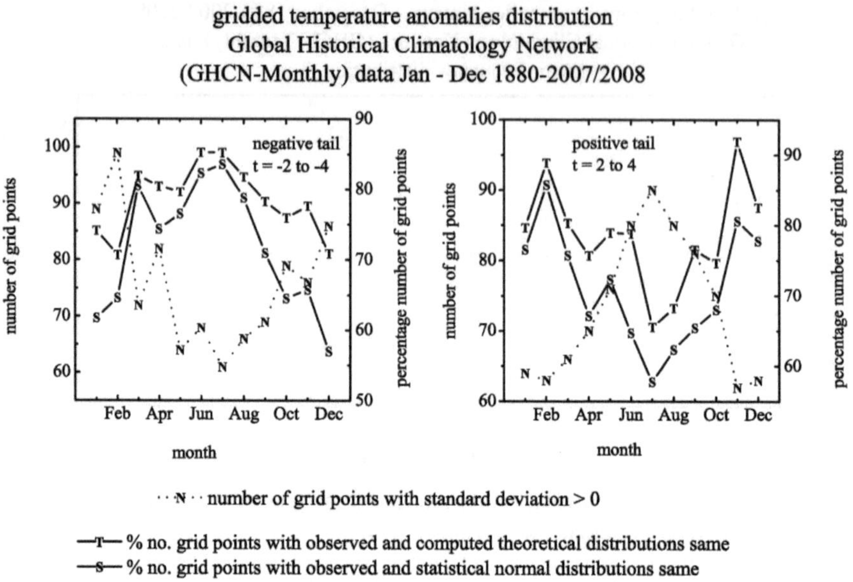

Fig. 5.5 The total and percentage numbers of observed extreme values same as model predicted theoretical and statistical normal distributions for positive and negative tail regions (normalized deviation t greater than 2) for the 12 months January to December

observed distribution included the computed theoretical (statistical normal) distribution within twice the standard deviation on either side of the mean then it was assumed to be the same as the computed theoretical (statistical normal) distribution at 5% level of significance within measurement errors. The number of observed distribution values which included the computed theoretical values and/or the statistical normal distribution values within twice the standard deviation on either side of the mean was determined. The total and percentage numbers of observed extreme values same as computed theoretical and statistical normal distributions are given in Fig. 5.5 for positive and negative tail regions (normalized deviation t greater than 2) for the 12 months January to December. The percentage number of observed extreme value points same as model predicted (computed) is more than the percentage number of extreme value points same as statistical normal distribution for all the 12 months. The observed distribution is closer to the model predicted theoretical than the statistical normal distribution.

5.4.2 *Continuous Periodogram Power Spectral Analyses*

The power spectra of frequency distribution of monthly mean data sets were computed accurately by an elementary, but very powerful method of analysis developed by Jenkinson (1977) which provides a quasi-continuous form of the classical periodogram allowing systematic allocation of the total variance and degrees of freedom of the data series to logarithmically spaced elements of the frequency range (0.5, 0). The cumulative percentage contribution to total variance was computed starting from the high frequency side of the spectrum. The power spectra were plotted as cumulative percentage contribution to total variance versus the normalized standard deviation t equal to $(\log L/\log T_{50}) - 1$ where L is the period in years and T_{50} is the period up to which the cumulative percentage contribution to total variance is equal to 50. The statistical *Chi-Square* test (Spiegel 1961) was applied to determine the 'goodness of fit' of variance spectra with statistical normal distribution which is close to model predicted variance spectrum for t values less than 2. The average and standard deviation of cumulative percentage contribution to total variance for each *normalized deviation t*(i) values were then computed for each month from the corresponding distributions for all the available grid points for the month. The average power spectra with corresponding standard deviations for each of the 12 months are plotted in Fig. 5.6. The total number of grid points and the percentage number of grid points with variance spectra same as statistical normal distribution are shown in Fig. 5.7. The mean power spectra follow closely the statistical normal distribution for the twelve months (Fig. 5.6). The power spectra mostly cover the region for *normalized deviation t* less than 2 where the model predicted theoretical distribution is close to the statistical normal distribution. A majority (more than 90%) of the power spectra

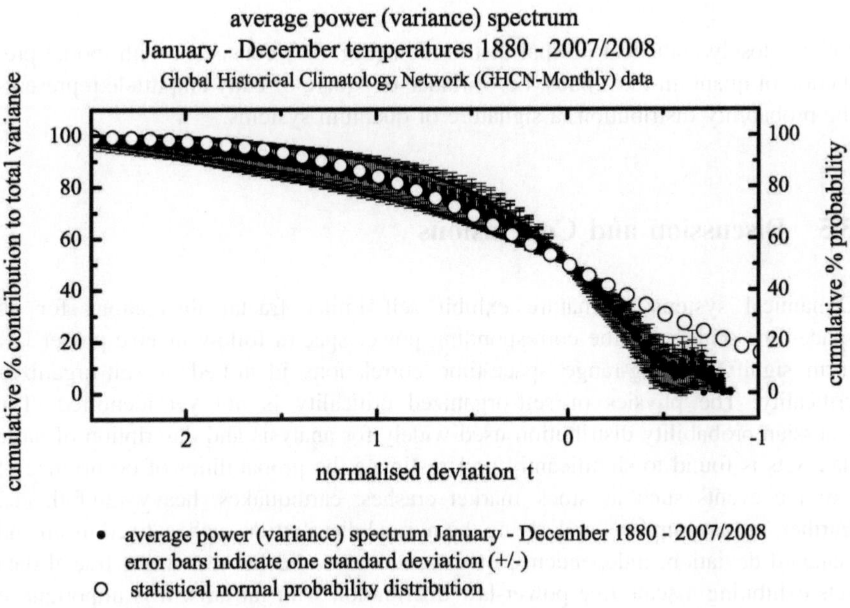

Fig. 5.6 Average power (variance) spectrum with vertical error bars indicating one standard deviation on either side of the mean. The statistical normal distribution is also shown in the figure for comparison

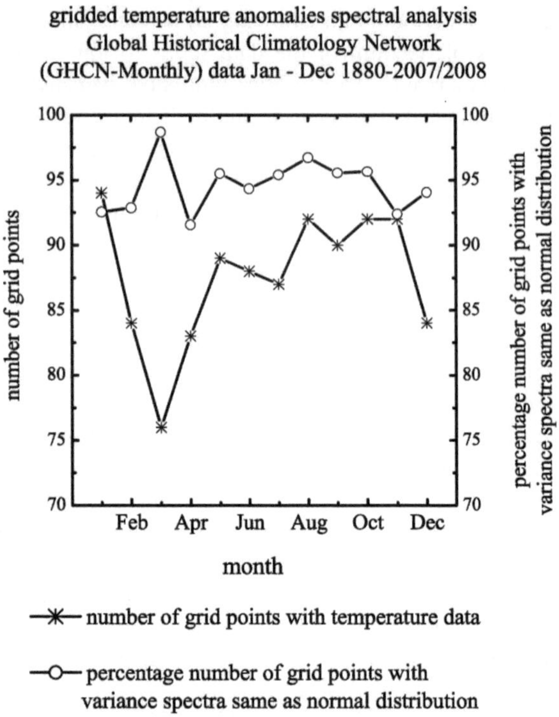

Fig. 5.7 The total number of grid points and the percentage number of grid points with variance spectra same as statistical normal distribution

follow closely statistical normal distribution (Fig. 5.7) consistent with model prediction of quantum-like chaos, i.e. variance or square of eddy amplitude represents the probability distribution, a signature of quantum systems.

5.5 Discussion and Conclusions

Dynamical systems in nature exhibit self-similar fractal fluctuations for all space-time scales and the corresponding power spectra follow inverse power law form signifying long-range space-time correlations identified as self-organized criticality. The physics of self-organized criticality is not yet identified. The Gaussian probability distribution used widely for analysis and description of large data sets is found to significantly underestimate the probabilities of occurrence of extreme events such as stock market crashes, earthquakes, heavy rainfall, etc. Further, the assumptions underlying the normal distribution such as fixed mean and standard deviation, independence of data, are not valid for real world fractal data sets exhibiting a scale-free power-law distribution with fat tails. It is important to identify and quantify the fractal distribution characteristics of dynamical systems for predictability studies.

A general systems theory for fractal space-time fluctuations (Selvam 1990, 2007, 2015; Selvam and Fadnavis 1998) shows that the larger scale fluctuation can be visualized to emerge from the space-time averaging of enclosed small scale fluctuations, thereby generating a hierarchy of self-similar fluctuations manifested as the observed eddy continuum in power spectral analyses of fractal fluctuations.

The model predictions are as follows:

- The probability distribution function P of fractal fluctuations follow inverse power law form τ^{-4t} where τ is the *golden mean*, and t, the normalized deviation is equal to $(x - av/sd)$ where av and sd are, respectively, the average and standard deviation of the distribution. The predicted distribution is close to the Gaussian distribution for small-scale fluctuations (normalized deviation t less than 2), but exhibits *fat long tail* for large-scale fluctuations (normalized deviation t more than 2) with higher probability of occurrence than predicted by Gaussian distribution. There is always a nonzero probability of occurrence of very large amplitude, damage causing extreme events.
- The inverse of the probability distribution function, i.e. $1/P$ equal to τ^{4t} represents the domain size extent of the internal fine scale (fractal) fluctuations of amplitude t (normalized deviation). High intensity extreme events corresponding to t values more than 2 occur with less probability over a larger size domain and are associated with widespread damage and loss such as in heavy rainfall, earthquakes, traffic jams, etc.
- The power spectra of fractal fluctuations also follow inverse power law form τ^{-4t} and t, the normalized deviation is expressed in terms of component periodicities L as $t = \{(\log L/\log T_{50}) - 1\}$ where T_{50} is the period up to which the cumulative percentage contribution to total variance is equal to 50. Since the power (variance, i.e. square of eddy amplitude) spectrum also represents the probability densities as in the case of quantum systems such as the electron or photon, fractal fluctuations exhibit quantum-like chaos.
- The precise geometry of the quasiperiodic Penrose tiling pattern underlie fractal fluctuations tracing out robust (self-sustaining) dominant bidirectional large eddy circulation $OR_OR_1R_2R_3R_4R_5$ (Fig. 3.1) associated with a factor of $20a^2$ $(20\tau^4)$ equal to 137.08203 increase in eddy energy flux expressed in terms of Feigenbaum's constant $\boldsymbol{a}$. This nondimensional dominant eddy energy flux is analogous to and almost equal to the *fine structure constant* $\propto^{-1}$ of atomic spectra; the energy spacing intervals between successive atomic spectral lines. Further, the energy increase between two successive fine structure eddy length step growths internal to the dominant large eddy domain may be expressed as a function of $(\boldsymbol{a}^2)^2$, i.e. proportional to the square of the *fine structure constant* $\propto^{-1}$. In the spectra of many atoms, what appears with coarse observations to be a single spectral line proves, with finer observation, to be a group of two or more closely spaced lines. The spacing of these fine-structure lines relative to the coarse spacing in the spectrum is proportional to the square of *fine structure constant*, for which reason this combination is called the *fine-structure constant* (Ford 1968).

Analysis of historic (1880–2008) data sets of global monthly mean temperature (GHCN V2) time series shows that the data follow closely, but not exactly the statistical normal distribution in the region of normalized deviations t less than 2 (Fig. 5.2). For normalized deviations t greater than 2, the data exhibit significantly larger probabilities as compared to the normal distribution and closer to the model predicted probability distribution (Figs. 5.3 and 5.4). A simple t test for 'goodness of fit' of the extreme values (normalized deviation $t > 2$) of the observed distribution with model predicted (theoretical) and also the statistical normal distribution shows that more number of data points exhibit significant (at 5% level) 'goodness of fit' with the model predicted (theoretical) distribution than with the normal distribution (Fig. 5.5). The mean power spectra follow closely the statistical normal distribution for the twelve months (Fig. 5.6). The power spectra mostly cover the region for *normalized deviation t* less than 2 where the model predicted theoretical distribution is close to the statistical normal distribution. A majority (more than 90%) of the power spectra follow closely statistical normal distribution (Fig. 5.7) consistent with model prediction of quantum-like chaos, i.e. variance or square of eddy amplitude represents the probability distribution, a signature of quantum systems. The model predicted and observed universal spectrum for interannual variability rules out linear secular trends in global monthly mean temperatures. Global warming results in intensification of fluctuations of all scales and manifested immediately in high frequency fluctuations. The general systems theory, originally developed for turbulent fluid flows, provides universal quantification of physics underlying fractal fluctuations and is applicable to all dynamical systems in nature independent of its physical, chemical, electrical, or any other intrinsic characteristic.

References

Bak, P.C., Tang, C., Wiesenfeld, K.: Self-organized criticality. Phys. Rev. A **38**, 364–374 (1988)

Feigenbaum, M.J.: Universal behavior in nonlinear systems. Los Alamos Sci. **1**, 4–27 (1980)

Ford, K.W.: Basic Physics. Blaisdell Publishing Company, Waltham, Massachusetts, USA (1968)

Grossing, G.: Quantum systems as order out of chaos phenomena. Il Nuovo Cimento **103B**, 497–510 (1989)

Jenkinson, A. F.: A powerful elementary method of spectral analysis for use with monthly, seasonal or annual meteorological time series. Meteorological Office, Branch Memorandum No. 57, London (1977)

Maddox, J.: Licence to slang Copenhagen? Nature **332**, 581 (1988)

Maddox, J.: Can quantum theory be understood? Nature **361**, 493 (1993)

Mandelbrot, B.B.: Les Objets Fractals: Forme. Hasard et Dimension, Flammarion, Paris (1975)

Peterson, T.C., Vose, R.S.: An overview of the global historical climatology network temperature database. Bull. Am. Meteorol. Soc. **78**, 2837–2849 (1997)

Peterson, T.C., Vose, R.S., Schmoyer, R., Razuvae, V.: Global Historical Climatology Network (GHCN) quality control of monthly temperature data. Int. J. Climatol. **18**, 1169–1179 (1998a)

Peterson, T.C., Karl, T.R., Jamason, P.F., Knight, R., Easterling, D.R.: The first difference method: maximizing station density for the calculation of long-term global temperature change. J. Geophys. Res. **103**, 25967–25974 (1998b)

Phillips, T.: The mathematical uncertainty principle. Monthly essays on mathematical topics November 2005, American Mathematical Society (2005). http://www.ams.org/featurecolumn/archive/uncertainty.html

Rae, A.: Quantum-Physics: illusion or reality?. Cambridge University Press, New York (1988)

Riley, K.F., Hobson, M.P., Bence, S.J.: Mathematical methods for physics and engineering, 3rd edn. Cambridge University Press, USA (2006)

Ruhla, C.: The physics of chance. Oxford University Press, Oxford (1992)

Schroeder, M.: Fractals, chaos and power-laws. W. H. Freeman and Co., N.Y. (1991)

Selvam, A.M.: Deterministic chaos, fractals and quantumlike mechanics in atmospheric flows. Can. J. Phys. **68**, 831–841 (1990). http://xxx.lanl.gov/html/physics/0010046

Selvam, A.M., Fadnavis, S.: Signatures of a universal spectrum for atmospheric inter-annual variability in some disparate climatic regimes. *Meteorology and Atmospheric Physics* **66**, 87–112. http://xxx.lanl.gov/abs/chao-dyn/9805028

Selvam, A.M., Fadnavis, S.: Superstrings, cantorian-fractal spacetime and quantum-like chaos in atmospheric flows. Chaos Solitons Fractals **10**, 1321–1334 (1999). http://xxx.lanl.gov/abs/chao-dyn/9806002

Selvam, A.M., Sen, D., Mody, S.M.S.: Critical fluctuations in daily incidence of acute myocardial infarction. Chaos Solitons Fractals **11**, 1175–1182 (2000). http://xxx.lanl.gov/abs/chao-dyn/9810017

Selvam, A.M.: Quantum-like chaos in prime number distribution and in turbulent fluid flows. Apeiron **8**, 29–64 (2001a). http://redshift.vif.com/JournalFiles/V08NO3PDF/V08N3SEL.PDF; http://xxx.lanl.gov/html/physics/0005067

Selvam, A.M.: Signatures of quantum-like chaos in spacing intervals of non-trivial Riemann zeta zeros and in turbulent fluid flows. Apeiron **8**, 10–40 (2001b). http://redshift.vif.com/JournalFiles/V08NO4PDF/V08N4SEL.PDF http://xxx.lanl.gov/html/physics/0102028

Selvam, A.M.: Cantorian fractal space-time fluctuations in turbulent fluid flows and the kinetic theory of gases. Apeiron **9**, 1–20 (2002a). http://redshift.vif.com/JournalFiles/V09NO2PDF/V09N2sel.PDF; http://xxx.lanl.gov/html/physics/9912035

Selvam, A.M.: Quantumlike chaos in the frequency distributions of the bases A, C, G, T in Drosophila DNA. Apeiron **9**, 103–148 (2002b). http://redshift.vif.com/JournalFiles/V09NO4PDF/V09N4sel.pdf; http://arxiv.org/html/physics/0210068

Selvam, A.M.: Quantumlike chaos in the frequency distributions of the bases A, C, G, T in human chromosome 1 DNA. Apeiron **11**, 134–146 (2004). http://redshift.vif.com/JournalFiles/V11NO3PDF/V11N3SEL.PDF; http://arxiv.org/html/physics/0211066

Selvam, A.M.: Chaotic climate dynamics. Luniver Press, UK (2007)

Selvam, A.M.: Rain formation in warm clouds: general systems theory. Springer Briefs in Meteorology, Springer (2015)

Spiegel, M.R.: Statistics. Schaum's Outline Series in Mathematics, McGraw-Hill (1961)

Index

A.M. Selvam, *Self-organized Criticality and Predictability in Atmospheric Flows*, Springer Atmospheric Sciences, DOI 10.1007/978-3-319-54546-2

MIX
Papier aus verantwortungsvollen Quellen
Paper from responsible sources
FSC® C105338

If you have any concerns about our products,
you can contact us on
ProductSafety@springernature.com

In case Publisher is established outside the EU,
the EU authorized representative is:
Springer Nature Customer Service Center GmbH
Europaplatz 3, 69115 Heidelberg, Germany

Printed by Libri Plureos GmbH
in Hamburg, Germany